CITIES, NATURE AND DEVELOPMENT

Sarah Dooling: To my parents and Marion
Gregory Simon: To Dimitri and Gabriel

Cities, Nature and Development
The Politics and Production of Urban Vulnerabilities

Edited by

SARAH DOOLING
University of Texas Austin, USA

and

GREGORY SIMON
University of Colorado Denver, USA

Routledge
Taylor & Francis Group

LONDON AND NEW YORK

First published 2012 by Ashgate Publishing

Published 2016 by Routledge
2 Park Square, Milton Park, Abingdon, Oxfordshire OX14 4RN
711 Third Avenue, New York, NY 10017, USA

First issued in paperback 2016

Routledge is an imprint of the Taylor & Francis Group, an informa business

British Library Cataloguing in Publication Data
Cities, nature and development : the politics and
production of urban vulnerabilities.
1. Urban ecology (Sociology) 2. Sustainable development.
I. Dooling, Sarah. II. Simon, Gregory.
307.7'6-dc22

Library of Congress Cataloging-in-Publication Data
Cities, nature and development : the politics and production of urban vulnerabilities / by Sarah Dooling and Gregory Simon, [editors].
    p. cm.
Includes bibliographical references and index.
ISBN 9781409408314 (hardback)
1. Urban ecology (Sociology) 2. Political ecology. 3. Environmental policy. 4. Nature-
-Effect of human beings on. I. Dooling, Sarah. II. Simon, Gregory, 1974-
HT241.C57 2011
307.76--dc23

2011021384

ISBN 13: 978-1-138-25536-4 (pbk)
ISBN 13: 978-1-4094-0831-4 (hbk)

Contents

PART 1 GEOGRAPHIES OF WEALTH AND RISK ACCUMULATION: NEOLIBERAL POLICY AND RESOURCE INSTRUMENTALISM

PART 2 UNANTICIPATED VULNERABILITIES: SUSTAINABILITY PLANNING, ENVIRONMENTAL MOVEMENTS, AND ACTIVISM

PART 3 VULNERABILITIES IN THE URBANIZING CONTEXT: CULTURAL AND DEMOGRAPHIC TRANSFORMATIONS

List of Figures

List of Tables

Acknowledgments

Sarah Dooling: many thanks to Elizabeth Mueller, University of Texas colleague and friend, for our work and discussions on equity and vulnerability. Thanks to my 2010 Urban Ecology seminar, where these ideas were discussed, challenged and refined. And lastly, thanks to Nishtha Meta, doctoral student, whose work has also been a springboard for developing these concepts.

Gregory Simon: I am grateful to members of the Lane Center and Spatial History Lab at Stanford University during various stages of this research, including Peter Alagona, Mathew Booker and Bob Wilson; and also Richard White for his guidance. A number of colleagues at CU Denver have provided immense support including Brian Page and Deb Thomas who so graciously shared her considerable insights on theories of vulnerability. For their thought provoking insights, I would like to thank members of the 2010 AAG panel "Conservation, Development and the Production of Urban Vulnerabilities" and the 2009 ACSP roundtable "People, Nature and Cities: New Directions for Urban Ecology in the 21st Century." Many friends and family have taken important supportive roles including Jesse Grant, Maria and Jeff Simon and the Helena Simon family. Thanks to Vicky for teaching me the meaning of hard work. And to Dimitri and Gabriel, you bring constant joy and inspiration to all my endeavors.

List of Contributors

Melissa Adams received her BA in Urban Studies from the University of Texas at Austin in 2009. She is currently a bilingual elementary school teacher in Austin. Her research interests center around bilingual mathematics education methods, race theory, particularly in terms of access to education.

Julian Agyeman is Professor and Chair of Urban and Environmental Policy and Planning at Tufts University. His research interests critically explore aspects of the complex and embedded relations between humans and the environment, whether mediated by institutions or social movement organizations, and the effects of this on public policy and planning processes and outcomes, particularly in relation to notions of justice and equity.

Timothy W. Collins is an Associate Professor of Geography in the Department of Sociology and Anthropology at the University of Texas at El Paso. He received his Ph.D. in Geography (2005) with a concentration in Urban Ecology from Arizona State University, where he was also a postdoctoral researcher (2005–2006) studying water resource conflict with the Decision Center for a Desert City. His research focuses on risks, hazards, disasters and urban environmental governance, and emphasizes issues of social vulnerability and environmental injustice in particular.

Sarah Dooling is Assistant Professor or Urban Ecology at the University of Texas in the School of Architecture and the Environmental Science Institute. Her research interests investigate the intersections between ecology and poverty, and ecology and violence.

Jessica K. Graybill is an Assistant Professor in the Geography Department at Colgate University. She holds a Ph.D. in Geography from the University of Washington with a concentration in Urban Ecology. Her ongoing research in the Russian Far East and in a variety of US urban settings focuses on human experiences with the environment, utilizing cognitive mapping and ethnographic field research techniques to elicit how people perceive and then use surrounding built and natural environments. Jessica works closely with indigenous and local peoples, as well as other stakeholders in government and corporate realms to present the kaleidoscope of meanings attributed to environmental use.

Anthony M. Jimenez is a Graduate Student and Teaching Assistant in the Department of Sociology and Anthropology, University of Texas at El Paso. He is in the process of earning his MA with foci in gender and international development. His current research investigates the underlying power dynamics within international development programs, specifically the role of gender ideologies. Additional research interests include neoliberalism, transnationalism, and globalization.

Kelvin Mason is a Senior Lecturer in the Graduate School of the Environment at the Centre for Alternative Technology in Machynlleth, Wales. With a Ph.D. in Human Geography, an MCs in Environmental Management and a first degree in Mechanical Engineering, his research interests range from social movements to building materials, coalescing around participation.

Harold A. Perkins is an Assistant Professor in the Department of Geography at Ohio University. He earned his Ph.D. from the University of Wisconsin-Milwaukee in 2006. His research interests include the political economy/ecology of urban environments, specifically the role of the state, governance, and voluntarism under neoliberalization.

Gregory Simon is an Assistant Professor at the University of Colorado Denver. He earned his Ph.D. in Geography at the University of Washington. He also spent two years as a Post-Doctoral Fellow at Stanford University. His research focuses broadly on complex systems of environmental governance. His projects aim to destabilize normative accounts of environment-development programs and expose unintended contradictions, new social-ecological narratives and alternative policy pathways.

Eliot M. Tretter received his Ph.D. from the Department of Geography and Environmental Engineering at the Johns Hopkins University in Baltimore, MD in 2004. He is currently a Lecturer in the Department of Geography and the Environment at University of Texas at Austin. His research interests have focused on the political economic dimensions of urban development and he has regional specialties in Northern Europe and the Southern United States. His articles have appeared in a number of urban studies and geography journals. He is presently working on a book called *Shadows of a Sunbelt City: Environmental Racism and Knowledge Economy in Austin, TX 1960–2000*. He would like to dedicate his manuscript to his late friend and mentor, the geomorphologist Gordan 'Reds' Wolman, who taught him everything he knows and loves about rivers.

Mark Whitehead is a Reader in Political Geography at the Institute of Geography and Earth Sciences, at Aberystwyth University, and Senior Research Fellow at the City Institute, York University, Toronto. He has written widely on urban and environmental politics and has a particular interest in the political geographies of nature and strategies for environmental behaviour change.

Part 1

Geographies of Wealth and Risk Accumulation: Neoliberal Policy and Resource Instrumentalism

Chapter 1

Cities, Nature and Development: The Politics and Production of Urban Vulnerabilities

Sarah Dooling and Gregory Simon

As the United States struggles with a persistent economic recession, concerns related to conditions and experiences of vulnerability have become the focus of academic scholarship and popular press coverage. With a US national unemployment rate hovering above 9 percent (Brookings Institute, 2010), many households are economically vulnerable, with some unable to retain their housing. In April 2010, e.g., one in every 45 US households received a foreclosure filing (Georgia Consumer Banking, 2011), up 16 percent from March 2009 (Tyrell, 2010). The financial struggles facing US households is mirrored internationally with the collapse of financial markets resulting in profound economic hardships for national governments, communities and households globally. In addition to the increasing economic vulnerability experienced by households worldwide over the past several years, changing environmental conditions – specifically changing climatic regimes – are also contributing to conditions and experiences of vulnerability. It appears as if the United States – and the world more generally – is facing unprecedented changes, as socio-economic stresses occur alongside environmental crises related to the degradation and depletion of usable land, water and air resources. Much of the scholarship and popular media coverage examining these conjoint economic and environmental crises has approached vulnerability as a state of being or condition measured at a particular point in time and experienced within a specific place.

Researchers working in disaster management and hazards, food and water security, and climate change fields have advanced theories that, collectively, help to define and advance the field of vulnerability studies (Turner et al., 2003; Kasperson ct al., 2005; Eakin and Luers, 2006; Ionescu et al., 2009). This expansive and progressive research arena has generally defined vulnerability as the degree to which a system (or series of interconnected systems) is susceptible and responsive to (either as adaptation or mitigation) the adverse effects of shocks and stresses (McCarthy et al., 2001). Assessing levels of exposure, sensitivity, and adaptive capacity to external pressures have provided insight into how conditions of vulnerability operate within particular populations (Wisner, 1993; Adger, 2006). Early studies concluded that conditions of overall vulnerability increase with a corresponding decrease in a given population's ability to prevent and recover from such stresses (Timmerman, 1981; Kates, 1985).

While these scholars focus on specific attributes of populations and their ability to respond to disasters and hardships, other scholars examine more explicitly the dynamics between pre-existing and emerging economic, environmental and social conditions that impact vulnerable communities. Liverman (1990) distinguished between the biophysical environment that contributes to vulnerability and the political, social, and economic conditions that also influence increased risk of exposure to harm. Wisner et al., (1994) linked increased levels of vulnerability to: a) hazardous environmental conditions and insufficient levels of natural resource availability, and b) the accessibility of provisioned services (e.g., health services, credit, information) and social resources (e.g., income, assets, familial support). Scholars have argued that the most rigorous and meaningful analyses of vulnerability are achieved from an interdisciplinary approach where considerations of social and ecological conditions are integrated (Cutter, 1996), even if dedicated analytic focus falls more heavily on natural systems, such as the mechanics of storm surges, or on social systems, such as low income housing policy.

Urban Vulnerability Analysis: From Static Condition to Dynamic Process

Scholars describing and measuring vulnerability have shifted their attention from largely rural contexts to urban settings. These studies have concluded that risks to urban environmental hazards are more complex phenomena, with overlapping risks associated with the household, workplace or neighborhood, and with uneven resource allocations and pollution risks from industrial contamination (e.g., Hardoy, Mitlin, and Satterthwaite, 2001). In urban and urbanizing environments, poverty and access to stable, affordable housing are key factors in determining a household's ability to withstand socio-economic stresses (Sanderson, 2000; Moser, 1998). Poverty alleviation and predictable and safe housing conditions enhance coping strategies for households responding to natural disasters, for it is the homeless and those without access to safe housing that are frequently most harmed by environmental hazards (Pelling, 2003). Along with poverty and housing access, the differential social impacts of local stresses resulting from, and contributing to, vulnerability are examined in the context of health, racial, gender and age composition of effected households and communities (Phillips et al., 2009).

However, experiences and conditions of urban vulnerability both result from and contribute to broader scale political, economic and environmental changes. For example, as Moser (1994) illustrates, the vulnerability of poor urban populations in developing countries must be understood within the context of global aid organizations by documenting the plight of communities affected by structural adjustment programs. Spatial scale becomes an important variable in understanding how conditions of vulnerability for urban populations are a response to, and a byproduct of, larger-scaled phenomena, including national policies, global financial markets, and regional environmental disasters. These studies reflect a shift from understanding vulnerability as measurable conditions towards conceptualizing vulnerabilities as conditions that are created and maintained through a series of historical relationships that interact across spatial scales. Here, researchers analyzing community responses to disasters and other destabilizing events, consider vulnerability as sets of dynamic conditions produced from historic interactions across economic, cultural, and social processes (Hewitt, 1987; Wisner et al. 1994; Pelling, 2003; Hogan and Marandola, 2005; Andrey and Jones, 2008).

The notion of vulnerability as multiple conditions that change through time is vastly different from the static "point-in-time" assessment approach which focuses on calculating rates of exposure, measuring risk-burden, and predicting impacts. The latter approach is characteristic of risk assessment and "vulnerability as outcome" studies (Kelly and Adger, 2000; O'Brien et al., 2007). In contrast, a more dynamic notion of vulnerability connotes how conditions of, and experiences with, vulnerability are produced through specific cross-scale interactions that are historical in nature. The authors in this book present research that draws from and expands upon the systemic and integrated socio-ecological drivers of vulnerability, i.e., in the *production of vulnerability*.

This volume follows the work of Findley (2005) who considers vulnerable spatialities to be the processes by which people and places are exposed to shifting states of vulnerability through a series of codified and enforced political economic agendas. Subsequent chapters in this volume similarly demonstrate that political ideologies and strategies influence access to human and natural resources for different social groups, thus designating responsibility for the production of vulnerability to politically and economically powerful institutions and individuals (Pelling, 2003). Conceptualizing vulnerability in terms of its historical production therefore allows researchers to identify the potential sources – including influential inter-agency and public-private alliances – that instigate, manage, and perpetuate vulnerable populations and places (Hogan and Marandola, 2005). A production-oriented framework provides conceptual space for analyzing how interactions between political economies of resource use and normative planning and management interventions – at both global and local scales – influence which places and populations are made vulnerable, and the intensity and persistence of conditions of vulnerability (Davis, 1998; Peet and Watts, 2004; Orsi, 2004; Wisner et al., 2004; Collins, 2005, 2008, 2010; Mustafa, 1998, 2005). Contributors to this volume develop thorough articulations of how conditions and experiences of vulnerability are produced, regulated, manipulated and resisted.

The Production of Urban Vulnerabilities: Key Analytic Themes

This volume theorizes the city, and its economic, cultural and environmental components through the lens of vulnerability. In the following pages we identify a suite of epistemological and ontological approaches that re-theorize urban nature as "material *and* narrated, ecological *and* political" (Braun, 2005, p. 642). Along with considerable work in the field of vulnerability studies, this volume builds on urban political ecological scholarship that emphasizes the mutually-constitutive relationship between economies, policies and ecological systems across temporal and spatial scales (Cronon, 1991; Gandy, 2002; Swyngedouw, 2004; Kaika, 2005). At the same time, various chapters seek to de-naturalize processes of urbanization and re-politicize the field of urban ecology by revealing and describing the deeply political and contentious processes through which urban ecological problems are produced, defined and mitigated (Keil and Desfor, 2004). A political urban ecological framework conceptualizes cities as complex integrated (economic, cultural and biophysical) systems that are governed through racial, gendered and class-based city politics, and undergirded by normative planning commitments to urban development, regional economic growth, and individual wealth accumulation (Wolch et al., 2002; Dooling et al,. 2006; Robbins, 2007; Brownlow, 2008). Contributions within this volume describe cities as socio-ecological arenas – at once produced through multi-scalar social and ecological processes and contested within spheres of formal and informal environmental politics (Heynan, et al., 2006). By explicitly developing the concept of *the production of urban vulnerabilities*, and by detailing its relationship to planning agendas that guide (oftentimes conjointly) urban sustainability, gentrification, suburban development, climate change adaptation, and other planning initiatives, the concept of urban vulnerabilities is a provocative conceptual lynchpin for the otherwise wide-ranging field of urban political ecology.

The contributing authors in this volume analyze a variety of urban vulnerabilities in North America and Europe that demonstrate two key points: (1) the dynamic nature and recursive processes involved in the creation, regulation, and manipulation of urban vulnerability and (2) practices of resistance to conditions and experiences of urban vulnerability as a form of political action. Several authors leverage dialectical analyses to reveal the concealed and invisible contradictions associated with sustainability related discourses, political movements and planning efforts (see chapters by Agyeman and Simons, Dooling, and Mason and Whitehead). The contradictions identified and articulated in these chapters challenge conventional notions that the benefits attributed to urban sustainability planning efforts are inherently positive, and that these benefits are distributed equitably and experienced uniformly. Some contributors conduct longitudinal analyses to demonstrate the effectual and affectual nature of vulnerability as it builds momentum over time (see chapter by Simon). In this chapter, a combination of urban economic development imperatives, real estate speculation and tax restructuring policies work alongside afforestation activities to produce vulnerability in a step-wise progression. Other contributors situate their analytic approach in urban political ecology frameworks that emphasize the impact of neoliberal economic strategies on cities struggling with significant demographic and industrial shifts (see chapter by Graybill); cities confronted by diverse and emerging urban park activities and preferences (see chapters by Perkins, Brownlow); and cities challenged by the disproportionate

North–South flow of environmental externalities stemming from First World risk offsetting behavior (see chapter by Collins and Jimenez). Two authors articulate how conditions of vulnerability are leveraged, manipulated and (for some) ignored in the political realm to support economic and ecological gains for political and economic elites (see chapters by Dooling, Perkins, Simon), while other authors demonstrate how racial prejudice works in concert with city planning efforts to disproportionately place minority and immigrant groups at risk to environmental hazards (see chapter by Tretter and Adams).

Resilience, as an analytic compliment to vulnerability, is explored by a number of contributors. Using historical analysis, Graybill documents the ways in which economic downturns in a rust belt city is counter-acted by proactively leveraging the economic and social capital associated with an influx of immigrants. Here resilience is conceived of as the capacity to respond positively to the condition of urban economic vulnerability and to avoid collapse due to industrial re-structuring. For Mason and Whitehead, resilience is a preparatory response to the threat of being vulnerable to the harms associated with climate change activism that values self-reliance in the face of an ineffective governmental response. Resilience can be conceived of as an individual's, neighborhood's, city's or social movement's capacity to resist persistent vulnerability, as well as the capacity to reverse the production of vulnerability.

For the editors of this volume, resilience and vulnerability are the two sides of the sustainability coin. Specifically, assessing vulnerability as a dynamic, temporally and spatially dependent phenomenon can lead to strategies for improving a population's or place's ability to resist or reverse vulnerable conditions, thus improving the capacity to adapt to, or mitigate, undesired future change absent significant disruption. Efforts intended to make cities sustainable that fail to consider processes generating future risks, will ultimately exacerbate vulnerabilities by deepening existing inequities and undermining the achievement of the very sustainability goals they propose to advance.

This edited volume demonstrates several analytic concepts related to urban vulnerability. First, vulnerability is more than a state of being to be assessed at a single point in time. Rather, it is generated through processes resulting from interactions between people (community, political and economic elites), institutional agendas (government agencies and non-profit groups), systems of production for goods and services (such as organic farms and global commodity markets) and government led planning strategies (including urban densification and suburban investments). The production of vulnerabilities also involves interrogating how resources and environmentally sensitive lands are managed and regulated (including floodplains, greenspaces, forests, areas of food production, location of polluting facilities) and how the physical terrain is constructed and designed (such as neighborhoods, streets and waterways). Second, the production of urban vulnerabilities can deepen existing harms while also creating new risks. Third, powerful alliances that direct the growth and development in cities can mobilize rhetoric of vulnerability in order to secure economic gain while perpetuating risks for already vulnerable people and places.

Fourth, experiences of vulnerability, driven by processes of political marginalization and economic segregation, can be mobilized into grassrooted political movements that expose, resist and transform the mechanisms that generate and reinforce conditions of being vulnerable. These mobilizations can potentially lead to the development of alternative planning and urban economic development trajectories.

This collection of essays does much more than use the condition of vulnerability itself as a means of explanation. Rather, it is vulnerability as a dynamic and recursive process that is thoroughly explained.

Conceptualizing Vulnerability: Six Clarifying Questions

We recognize that the dynamics involved in producing and encountering vulnerability are complex and traverse spatial and temporal scales, i.e., the mechanisms that facilitate the production of urban vulnerabilities, and the diverse responses among those at risk, are context specific. The chapters in this book also reflect a placed-based, constructivist approach to conceptualizing vulnerability, where theories about urban vulnerability are developed and refined using empirical data from specific locales. By focusing on local histories, politics and biophysical characteristics, chapters in this volume articulate how the production of urban vulnerabilities is mediated locally in the face of larger-scale processes.

We pose a series of questions to clarify nuances and complexities associated with the volume's constructivist, context-dependent conceptualization of vulnerability.

First, given that conventional research has typically concerned itself with static, point-in-time assessments, our first question asks: *what does it mean to be vulnerable?* Drawing from chapters in this volume, conditions of vulnerability include: being impoverished, being an immigrant and lacking predictable, safe housing (Dooling); owning property in areas with high fire loads stemming from a century of localized wealth and biomass accumulation (Simon); being African American or Mexican living in flood prone areas (Tretter and Adams); residing in a city that has lost its industrial base (Graybill); and lacking access to affordable, organic food that reflects culturally specific ingredients and produce (Mason and Agyeman).

A logical next question to ask is: *what are the (potential and actual) consequences of being vulnerable?* Drawing from the work in this volume, the consequences of codified, racially discriminatory city plans has resulted in African Americans disproportionately living in urban areas prone to flooding, thus increasing their risk of property damage and displacement (Tretter and Adams). Low-income immigrant households in a neighborhood targeted for transit-oriented development face an increased risk of being displaced due to rising housing costs and the demolition of affordable units resulting from neighborhood improvement initiatives (Dooling). Activist groups calling attention to urban vulnerabilities related to climate change themselves become vulnerable to the psychological and physical dangers of demonstrating in the public realm (Mason and Whitehead). Long-term city disinvestment in public amenities, such as parks and green spaces, results in places where residents feel unsafe, and where crime is perceived to be high. Conditions of "produced" vulnerability are then used by city managers to justify the allocation of city funding to maintain parks in wealthier, white neighborhoods (Brownlow, Perkins). Other consequences for parks that are vulnerable to declining city funds include the replacement of city employees with park volunteer labor, which ultimately results in lost jobs (Perkins). For cities, the consequence of being vulnerable to furthering economic decline entails developing political–economic strategies that promote the city as a welcoming and economically viable place for immigrants (Graybill). This volume presents numerous ways of experiencing vulnerability and argues that the intensity and duration of being vulnerable – which may include health, economic and quality of life indicators – vary within and across populations and places.

While these examples illustrate connections between conditions and consequences of being vulnerable, there are additional questions related to identifying multiple vulnerabilities connected to the same event, region or set of urban conditions. Chapters in this book thus ask the following question: *how are multiplsase vulnerabilities connected – not only in relation to the originating event or place, but also in relation to each other?* How might researchers analyze such complexity? The analysis of interacting vulnerabilities requires a process-oriented conceptualization of vulnerability and benefits from a dialectical methodological approach. Dialectical analyses move beyond the identification of feedback loops that are part of systems theory. While feedback loops identify both reinforcing (negative) and destabilizing (positive) influences among system variables, they do not necessarily facilitate, nor are conventionally used in, the identification of how multiple feedback loops relate to each other. This is an important distinction as dialectical analysis focuses on the structure of processes (rather than objects or conditions themselves) that involves multiple spatial scales and local histories (Lewontin and Levins, 2007).

Dialectical analysis also involves analyzing contradictions that contribute to and emerge from interacting feedback loops in complex urban environments. Identifying contradictions is valuable for being able to predict unintentional outcomes associated with urbanization, development and conservation efforts; and for providing insight into how conservation and sustainability related efforts are undermined by alliances and tactics that are not easily observable or recognizable. In this volume, Dooling discusses how goals of increasing public ridership and lowering the city's carbon footprint through transit-oriented development is undermined by displacing already public-transit dependent households to the urban fringe lacking public transportation options, which will lead to future increases in the private vehicle miles traveled (VMTs) for the region.

Other contributors to this volume reveal how risk and exposure to harm associated with a single phenomenon is experienced differently across populations and places. Simon describes how Oakland Hills residents in fire prone areas are vulnerable to periodic conflagrations; yet, residents living in the flatlands far removed from frequent fire disasters experience higher net vulnerability. The poorer populations located in the flatlands have reduced access to insurance risk-offsetting resources, and are thus more likely to incur the full extent of costs associated with fires. Actively incorporating first and second order analysis of vulnerability can yield considerable analytic dividends. Simon reveals that while exclusive neighborhoods are constructed in high-risk fire areas, their attempts to maintain an economically exclusive neighborhood through tax reforms had the secondary effect of exacerbating vulnerability in their immediate neighborhood and in the Oakland flatland areas where poor and minority people lived as a result of depleted fire department budgets. As these examples suggest, this volume argues that understanding conditions of vulnerability in the fullest sense requires a relational approach in order to account for social–ecological feedbacks, the interactions among destabilizing and reinforcing feedback loops, and multiple forms of vulnerability within a single analysis.

A third clarifying question relates to the persistence of being at risk, and how the intensity and duration of risk varies through time. *Does being vulnerable mean constant endangerment, or does it imply existing in a state of elevated, but not prevailing, risk?* Tretter and Adams detail the fluctuating risk African Americans and Mexican immigrants experience to their homes being flooded as a result of racial policies that allocated their settlements to environmentally hazardous parts of the city. Experiences of being at risk can intensify as conditions change and threats emerge; likewise, experiences of risk can be alleviated by minimizing exposure to harm and shifting the risk to a different place or group of people. Collins and Jiminez describe various forms of vulnerability associated with the North–South transfer of harmful environmental externalities associated with neoliberal economic policies. Some populations are made vulnerable through community displacement practices that push agrarian communities into urban slums. Other communities are forced to reside near manufacturing and waste facilities that poison airsheds and adjacent water bodies. With these developments, some urban residents will be consistently vulnerable with a relatively low risk of morbidity (living in slums with constant health risks associated with poor sanitation, malnutrition and disease), while others are exposed to conditions that are more immediately life-threatening such as high exposure to carcinogenic dioxins from waste incineration. Mason and Whitehead illustrate the persistent risks associated with climate change that, in turn, spur various political activists to challenge those perceived to be responsible for their endangerment. Yet, these individuals themselves become subjected to periodic (i.e., less persistent) but considerably more acute forms of vulnerability for speaking out on the issue. These authors demonstrate that the duration and intensity of risk, with its associated condition of vulnerability, fluctuates depending on what people are at risk to – poverty, toxins, or violence. While the magnitude of impacts is a crucial variable for defining vulnerability, so too is the frequency of exposure to high-risk conditions. Thus, developing a robust conceptualization of vulnerability in the context of uneven exposure over time requires establishing explicit temporal parameters to guide analysis.

The fourth clarifying question explores the extent to which conditions of vulnerability are embedded within deeply rooted hierarchical, political, economic and social relations along lines of income, race, gender and citizen status. *How are conditions of being vulnerable (to flooding, displacement, and wage loss) created and perpetuated by highly uneven levels of access to economic resources, political power and strategic alliances?* Dooling's work in this volume sets apart two distinct, albeit interacting, lines of enquiry: (1) what are the sustainability planning processes through which politically and economically marginalized populations are put at risk? And (2) how does the experience of being persistently at risk contribute to a deepening sense of being disempowered for those marginalized within public planning processes? These lines of enquiry connote a difference between marginalized populations being made vulnerable by the effects of historical patterns of disinvestment, and the marginalizing influence of vulnerability, where the latter highlights the regressive momentum (i.e., increased disempowerment) that may accompany and reinforce experiences of living under elevated levels of risk. Dooling argues that patterns of long-term disinvestment in a poor, immigrant neighborhood reflect city-wide strategies to concentrate affordable housing in these communities which, in turn, contributes to the area's overall susceptibility for continued experiences of substandard housing quality, crime, and flooding. The pattern of disinvestment, and the resulting concentration of primarily immigrant, low-income, rental households, contributes to the neighborhood's relative lack of political leverage in city planning efforts, which contributes to a sense that the low-income neighborhood residents are excluded from meaningful involvement in decisions about the neighborhood's future.

Contributors to this volume also illustrate how conditions of vulnerability are exploited to achieve economic and political gains for city elites. Collins and Jimenez argue that a political economic analysis of vulnerability acknowledges the influence of policies that produce on-the-ground conditions of vulnerability for poor people while simultaneously concentrating wealth for elites. Perkins argues that constricting city budgets for parks management is used to justify the expansion of volunteerism and the laying off of city park employees, thus exploiting the city's budgetary vulnerability to alternative employment practices that maintain fiscal solvency. Additionally, exploitation of vulnerability can result in the allocation of funds for securing and maintaining resources and amenities for wealthier segments of urban populations. Perkins demonstrates that in order to develop commodity parks for wealthy urban clientele, the city of Milwaukee intentionally neglected parks through many years of disinvestments in maintenance and recreation programs. More exclusive, privately funded park spaces with expensive restaurants have been created, in part by making the parks previously used by the more impoverished segments of society less desirable, and even dangerous.

Agyeman and Simons describe how the identification of healthy food deserts spurred the creation and spread of alternative food networks. Yet locavore movement boosters have done surprisingly little to bring culturally appropriate local organic foods to poor neighborhood communities and, instead, generated considerable social and financial resources for securing and increasing healthy foods options for middle and upper class white residents. In the case of local foods and city parks, community disparities widen, and not necessarily because less food or open space is available for certain communities, but rather because the status of the resources they use – as inadequate and in need of improvement – was leveraged to secure new and upgraded resources for more privileged members of society.

These chapters highlight the relationship between material and symbolic vulnerabilities. They show that in many situations, addressing the plight of vulnerable communities is less important than mobilizing public concern for them, and that their instrumentalist mobilization ultimately advances the goals and planning objectives of elite community members. It can be instructive, then, to view vulnerability for certain communities *in relation* to vulnerability's counterpart: the augmentation of privilege, and access to desirable and essential resources for more resilient segments of society. Exploiting the material and symbolic conditions of vulnerability in a manner that benefits certain groups, ultimately widens disparities and undermines efforts to create socially and environmentally just places.

The disempowering impacts associated with political and economic exploitation of vulnerabilities are crucial to our overall conceptualization of vulnerability. Equally valuable, however, are questions related to the possibilities for resisting and transforming experiences of being vulnerable. A fifth clarifying question asks: *how do communities mobilize to transform conditions of vulnerability?* More than a social–ecological outcome, the concept of vulnerability provides analytic space for the careful consideration of societal responses and political movements. Various chapters discuss political and social mobilizations that actively resist being vulnerable through efforts that aim to transform governance systems and policies. Contributors to this volume articulate urban vulnerabilities as starting points, and not ending points, in how cities are planned and developed. Vulnerable urban populations are both acted upon and activated, as a source of development outcomes *and* starting points. As Gibson-Graham (2006) notes, the project of examining hegemonic formations that generate social vulnerability must also include efforts "to contemplate its destabilization" and imagine individuals and communities as "'made' and 'as making' themselves" (2006, p. 23). While some social groups may reinforce dominant planning agendas, this volume demonstrates how other groups, from park users to climate change activists, may chart new urban development trajectories.

Brownlow documents how African American women self-organized to restore a neglected urban park and transform the park into a valued public amenity. He argues that the motivations for volunteering are political and personal, derived from histories of racial discrimination and political neglect. The women's restoration efforts accomplish many goals, including reversing conditions of environmental degradation, resisting the historical marginalization of African American women in the city, and enhancing the construction of an "insurgent citizenship" (Holston, 1999). Mason and Whitehead focus on a political and cultural movement associated with climate change politics known as the Transition Culture. These community groups expose how cities are vulnerable to the impact of peak oil and climate change; Transition Culture emphasizes local self-reliant strategies that build internal capacities (resilience) to resist harms associated with future risk. Meanwhile, Graybill describes the cross-scale strategies among city boosters and state and federal governmental agencies to revitalize a Rust Belt city now void of its industrial economic base and faced with a recent influx of immigrants. Taken collectively, these chapters demonstrate the various strategies of social mobilization, including: (1) community groups that effect change within existing governmental structures (Brownlow); (2) activist groups that are suspicious of government's effectiveness and self-organize outside of governmental support (Mason and Whitehead); and (3) city-wide co-ordination of multiple government agencies to develop larger-scaled responses to economic and social change (Graybill).

The sixth clarifying question points to the heart of this volume, whereby we ask: *how are urban vulnerabilities produced?* How do various existing conditions of vulnerabilities, inequities, structural hierarchies and neoliberal practices and government policies contribute to the production of vulnerabilities? What are the implications of conceptualizing vulnerability as a process oriented phenomenon? Each contributor to this volume demonstrates how conditions and experiences of being vulnerable are built upon and out of historical alliances, urban inequities, and political economies of wealth accumulation and patterns of dispossession, and marginalization. When placed in a historical perspective, the production of vulnerability, by virtue of its temporal aspects, includes strategies of maintenance, regulation, exploitation and – in some instances – resistance. In this way, we consider vulnerability as not necessarily an end-point condition, but an urban phenomenon that dynamically shifts intensity and duration through time and across space, and from which contradictions emerge that are frequently unarticulated. The key intellectual contribution of this volume is demonstrating the diversity of mechanisms and processes through which urban vulnerabilities are produced, the emerging contradictions, and the implications this has for sustainable and resilience urban development. The chapters that follow are organized into categories of production, and are briefly summarized below.

Production of Vulnerabilities: Analytic Themes in Chapters

Geographies of Wealth and Risk Accumulation: Neoliberal Policy and Resource Instrumentalism

Simon frames his historical analysis of the 1991 Oakland Hills Firestorm (Oakland, CA) event by highlighting how household fire risk was founded upon and reinforced by a broader regional political economic and environmental history that was premised on an unremitting commitment to wealth accumulation and instrumentalist land use policy. Vulnerability, defined as the risk of harm and exposure to fire, as well as the burden of purchasing fire insurance for not-wealthy residents, is produced through early resource extraction and property speculation activities closely linked to the development of San Francisco Bay Area townships as well as emerging suburban conservative homeowner politics, subsequent statewide tax restructuring policies, and uneven risk offsetting resources. Simon makes an important contribution to the study of urban vulnerability by clearly articulating how conditions of vulnerability become more acute and chronic over time, first as a state of *effect* stemming from regional development and resource use policies, and second a state of *affect* that engenders further land use responses and produces new and enhanced levels of vulnerability. In this way, vulnerability is never produced as a planning outcome only. Instead, conditions and experiences of vulnerability intensify and persist as they compound and gain momentum over time. After nearly 150 years of vulnerability in production, Simon's chapter concludes by describing how those most implicated in property and wealth accumulation activities are buffered from substantive fire risk, while poorer members of Oakland's flatlands – far removed from the region's instrumentalist land use policies – experience the fullest risk burden.

Marginalization, and its corresponding theme of facilitation, is explored by Collins and Jimenez. These authors conceptualize vulnerabilities in the context of processes that displace and transfer risk; as economically powerful communities and regions externalize risks onto poorer and less powerful populations. Risk displacement involves the simultaneous processes through which risk is created (facilitation) and received (marginalization). The authors outline three ways in which unequal risk and vulnerability are produced under a neoliberal economic order. First, the emergence of disaster capitalism, enabled by neoliberal institutional arrangements, allows political and economic elites to transfer risks to less powerful communities while expropriating rewards stemming from socionatural disasters. Second, technological risks are transferred to from the global North to South as wealthy states and populations are permitted to accumulate capital while displacing the deleterious and toxic wastes of their consumption activities onto less powerful communities (who then must cope with harmful exposures). Lastly, with land privatization and decreasing funds for social programs, urban in-migration has increased with the concurrent proliferation of slums, resulting in more people being exposed to hazardous living conditions. These three modalities of uneven risk production lead the authors to pivot from theories of "accumulation by dispossession" to a conceptual ordering around the notion of "accumulation by endangerment."

Two contributors focus on the ways in which public urban parks are vulnerable to budgetary constraints. These chapters examine the implications of municipal strategies taken to address insufficient funds for park management within neoliberal modes of governance. Perkins describes market-oriented urban environmental governance strategies that organize park patronage around public–private sector partnerships that prioritize market profitability and personal responsibility. Focusing on Milwaukee, WI, Perkins details how disinvestment in many city parks during the last thirty years reduced patronage and closed facilities, leading citizens to think of parks as increasingly dangerous places to visit. Meanwhile the emergence of commodity parks exemplifies a shift in the management of urban nature, as parks are no longer invested for the sake of providing urban residents recreation and rejuvenation in green space. Poor and minority citizens, in particular, are susceptible to a compromised quality of urban life when active and passive recreational opportunities in safe park spaces become dependent on the park's ability to participate in these market practices. Volunteering in specific parks is used to rationalize reducing unionized municipal park caretakers. The net effects of a for-profit parks management system are lost wages and health benefits for many employees, which in turn, jeopardize their ability to provide for themselves and their families.

Unanticipated Vulnerabilities: Sustainability Planning, Environmental Movements, and Activism

Authors in this section document unanticipated vulnerabilities that are produced through sustainability planning efforts, environmental movements that promote locally grown organic food, and activism that calls attention to the dangers of climate change. Agyeman and Simons provide an overview of the emergence of the locavore movement, which seeks to generate local food systems, ecologically sensitive production techniques, and locally oriented economic development. This has resulted in a proliferation of urban farmer's markets and Community Supported Agriculture (CSA) projects that provide a more holistic and direct connection between food producers, consumers and the agricultural landscape (urban and rural). Whether perceived as an anti-globalization effort, rural economic development initiative or as a component of broader urban sustainability programs, the local organic food movement has been touted as an important strategy for creating healthy, ecologically sustainable and socially just cities. Drawing from environmental justice literature, Agyeman and Simon challenge these assertions by demonstrating how local food access is structured along race and class lines. The authors demonstrate that access to local organic food, in the form of farmer's markets, is concentrated in middle to upper income neighborhoods, while poorer neighborhoods – mostly communities of color – exist in conditions of food insecurity, where food access is limited to nutritionally unhealthy options, and where risk to associated health impacts is high. Analyzing access to locally, organically produced food along race and class lines reveals how access to such food for those most in need fails to materialize, and that poor and minority community members continue to experience disproportionately higher levels of risk to poor nutrition and food insecurity.

Dooling exposes other contradictions associated with planning efforts intended
to enhance urban sustainability by exploring cases of ecological gentrification
in Seattle, WA and Austin, TX. The Seattle case demonstrates how designating
public green spaces for ecological purposes – including carbon sequestration,
habitat connectivity and expansion of pervious cover – results in the expulsion,
banishment and, in some cases, arrest of homeless people camping in these spaces.
In seeking to reduce their vulnerability to crime and disease associated with
shelters and temporary housing, homeless individuals increase their vulnerability
to different risks – expulsion from green spaces and (for repeat offenders) arrest.
Whereas designating green spaces for ecological purposes allows, and even
promotes, the access to for non-profit and educational groups, the vulnerability
of homeless people is both ignored ("it's not our problem" attitude among park
planners) and exacerbated (through expulsion and arrest). The enforcement of
civility ordinances, which are intended to regulate public behavior of those people
who enact their private lives publically, become mechanisms in the production
of social vulnerabilities in the context of sustainability planning. The Austin case
demonstrates that both social and ecological vulnerabilities can be produced
from sustainable transportation projects intended to enhance urban ecological
functioning. The approved transit-oriented development plan situates low-income
immigrant rental households' citizens vulnerable to future displacement through
inadequate production and preservation of affordable rental housing options. With
some households relocating to the urban fringe where public transit is lacking, the
miles traveled using private transportation by this previously transit-dependent
population will most likely increase, which will likely undermine the sustainability
goals articulated in the plan. Dooling demonstrates that ecological change is never
socially neutral, that multi-scalar analyses are critical for identifying social costs
and vulnerabilities that are produced by ecologically driven planning efforts.

Drawing upon a Lefebvrian framework, Mason and Whitehead focus on how
the mobilization of a politics of vulnerability is associated with three modalities
of vulnerability that co-exist in the urban context: (1) vulnerabilities that are
produced beyond the city-scale yet directly impact the urban environment (e.g.,
climate change); (2) perceptions of vulnerabilities that are constructed by various
groups in order to serve political agendas; and (3) experiences of being physically
and psychologically vulnerable that are associated with being a political activist.
The Transition Culture movement, a community-based mobilization, leverages
media framings of climate change and peak oil threats to the future of cities. The
movement proposes a planned strategy for energy descent based on re-organizing
urban economies around smaller metropolitan areas. The authors describe how an
activist camp, by virtue of its anti-capitalism stance, is vulnerable to the coercive
forces of the state (i.e., arrest and police violence) as well as to an unsympathetic
(and occasionally violent) public that create stress and psychological trauma for
activists. The authors conclude that the use of vulnerability as progressive political
tool produces new kinds of urban anxiety and potential trauma, and the challenge
for activists is to establish the real and present danger of urban vulnerabilities to

climate change while minimizing feelings of trauma that could potentially inhibit radical political action. This paper provides insights into the interactions between structural effects and individual affects associated with the production of urban vulnerabilities.

Vulnerabilities in the Urbanizing Context: Cultural and Demographic Transformations

These authors explore the production of urban vulnerabilities through nuanced analyses that considers historical demographic transformations. In contrast to Perkins chapter on parks (see previously), Brownlow focuses on how a group of African American women in Philadelphia, PA self-organized to restore a neglected urban park in their neighborhood. Their initiative and persistent commitment to the park's ecological restoration emerged in the face of persistent neglect of parks in minority neighborhoods on the part of the city parks department and the resulting self-imposed exile of women from these parks due to unsafe environments. The women self-organize based on experiences of being vulnerable to continuing park department policies of neglect that is perceived as a form of injustice. Whereas Perkins frames volunteerism as producing harm and risk for city employees, Brownlow describes volunteerism as opening up a political space for women who have experienced racial and economic prejudice and as an opportunity for (a temporary and) safe re-entry into a historically important space for community gatherings and recreation. In addition, the women's restoration effort occurs within state sanctioned policies and discourses. Drawing from feminist analyses of women's re-appropriation and politics of place, Brownlow argues that, in Philadelphia, volunteerism calls attention to and ultimately challenges racialized histories of injustice and neglect and corresponding conditions of vulnerability. Thus, the women's volunteer restoration effort works to resist future neglect, spatial exclusion, environmental degradation and the production of unsafe urban parks. While Brownlow's analysis points to the impact of neoliberal policies related to volunteerism, his primary contribution to vulnerability studies focuses on integrating feminist and urban political ecology theories to reveal and challenge women's marginalization within and resistance to urban geographies marked by racism and classism.

Graybill reveals the ways in which re-articulating and re-aligning processes associated with producing vulnerabilities can lead to strategies that enhance ecological function, economic productivity and new potentials for urban vitality. Focusing on the rust belt city of Utica, NY, Graybill's historical narrative documents the forces that have contributed to ecological and economic conditions of vulnerability: abandoned brownfield sites, declining city population, outmigration of the creative class to the suburbs, declining tax base, and shifting cultural identities among generations of immigrants. She demonstrates how reframing conditions of vulnerability as a potential basis for revitalization can facilitate city strategies, including securing federal and local resources for refugee resettlement in the wake of precipitous population decline. At the urban scale, the city addressed two conditions of vulnerability (massive population decline and long-term tax base decline) by attracting and supporting new immigrant communities. Looking to the future, abandoned properties and areas containing a high number of brownfield sites can be places of future urban in-fill development. The major conceptual contribution of this work to vulnerability studies is the framing of urban development as a series of interactions between vulnerabilities and resilience, and that the interplay between these conditions are generated, managed and abated by influential governmental, and local actors operating across scales.

Tretter and Adams demonstrate how the valuation of land based on vulnerability to natural hazards (i.e., flooding) has historically intersected with policies of white supremacy in order to organize the shifting historical geographies of race and class during the Jim Crow era in Austin, TX. Drawing from environmental racism literature, these authors describe the history of urban policies that sanctioned practices of segregation for African Americans and also Mexicans immigrants. For these two minority groups, affordable land was located in floodplains. Vulnerability – defined as the disproportionate risk to flooding – is produced through city policies and urban land economics, which in concert with segregationist politics, is transferred along lines of race and class. Understanding the dynamics of transferring the burden of risk, and thus the experience of being vulnerable, necessitates an understanding of racial politics, of not only African Americans but also, in Austin, Hispanics. As the authors note, both of these groups do not receive the privilege of staying dry, and the burden of flood risk is produced through the confluence of segregationist politics and land use economics. These authors major contribution to vulnerability studies is their application of an environmental justice framework to an historical analysis of the production of vulnerability to urban flooding.

All of the chapters following this volume may be organized around two primary meta-objectives. First, this book uses *vulnerability as an epistemological tool* to explain and demonstrate the magnitude, frequency, distribution, and directionality of vulnerability within particular places and populations, and also to explain emerging and persistent structural processes and practices that produce vulnerability over time and space – ranging from neoliberal economic policy to regional suburbanization and urban sustainable planning initiatives. This book describes a number of dialectical relationships to illustrate these and various other modalities of vulnerability's production. Second, this book uses a series of case studies to articulate the inner mechanisms and dynamics involved in producing vulnerabilities. Chapters in this volume will assist in explaining *the ontology of vulnerability* as a complex process that fluctuates through time and across space; that generates contradictions between intentions and outcomes; that increases in intensity, gains momentum, reinforces and transfers conditions of vulnerability; and that dissipates, undermines, and challenges those very same conditions.

References

Adger, N. (2006). Vulnerability. *Global Environmental Change* 16(3): 268–281.

Andrey, J. and Jones, B. (2008). The dynamic nature of social disadvantage: implications for hazard exposure and vulnerability in Greater Vancouver. *The Canadian Geographer* 52: 146–168.

Blaikie, P. and Brookfield, H. (1987). *Land Degradation and Society*. London: Routledge.

Braun, B. (2005). Writing a more-than-human urban geography. *Progress in Human Geography* 29: 635–650.

Brookings Institute. (2010). Metromonitor: tracking economic recession and recovery in America's 100 largest metropolitan areas. Retrieved from: <http://www.brookings.edu/~/media/Files/Programs/Metro/metro_monitor/2010_09_metro_monitor/0915_metro_monitor.pdf>.

Brownlow, A. (2008). *A Political Ecology of Neglect: Race, Gender, and Environmental Change in Philadelphia*. Saarbrucken: Verlag DM

Collins, T. (2005). Households, forests, and fire hazard vulnerability in the American West: a case study of a California community. *Global Environmental Change B: Environmental Hazards* 6(1): 23–37.

Collins, T. (2008). The political ecology of hazard vulnerability: marginalization, facilitation and the production of differential risk to urban wildfires in Arizona's White Mountains. *Journal of Political Ecology* 15: 21–43.

Collins, T. (2010). Marginalization, facilitation, and the production of unequal risk: the 2006 Paso del Norte floods. *Antipode* 42(2): 258–288.

Cronon, W. (1991). *Nature's Metropolis: Chicago and the Great West*. New York: W.W. Norton.

Cutter, S. (1996). Vulnerability to environmental hazards. *Progress in Human Geography* 20(4): 529–539.

Cutter, S.L., Mitchell, J.T., and Scott, M.S. (2000). Revealing the vulnerability of people and places: a case study of Georgetown County, South Carolina. *Annals of the Association of American Geographers* 90(4): 713–737.

Davis, M. (1998). *Ecology of Fear: Los Angeles and the Imagination of Disaster*. New York: Metropolitan Books.

Dooling. S., Simon, G., and Yocom, K. (2006). Place-based urban ecology: a century of park planning in Seattle. *Urban Ecosystems* 9(4): 299–321.

Eakin, H. and Luers, A. (2006). Assessing the vulnerability of social-environmental systems. *Annual Review of Environment and Resources* 31: 365–394.

Findlay, A.M. (2005). Editorial: vulnerable spatialities. *Population, Space and Place* 11: 429–439.

Gandy, M. (2002). *Concrete and Clay: Reworking Nature in New York City*. Cambridge: MIT Press.

Georgia Consumer Banking (2011). Accessed from: <http://georgiaconsumerbanking.com/2011/01/13/national-foreclosures-to-rise-in-2011.html>.

Hardoy, J.E., Mitlin, D., and Satterthwaite, D. (2001). *Environmental Problems in an Urbanizing World*. London: Earthscan.

Harvey, D. (1996). *Justice, Nature and the Geography of Difference*. Oxford: Blackwell Publishers.

Hewitt, K. (1997). *Regions of Risk. A Geographical Introduction to Disasters*. Essex: Addison Wesley Longman.

Heynen, N., Kaika, M., and Swyngedouw, E. (2006). *In the Nature of Cities: Urban Political Ecology and the Politics of Urban Metabolism*. New York: Routledge.

Holston, J. (1999). *Insurgent Citizenship: Disjunctions of Democracy and Modernity in Brazil*. Princeton: Princeton University Press.

Ionescu, C., Klein, R.J.T., Hinkel, J., Kumar, K.S.K., and Klein, R. (2009). Towards a formal framework of vulnerability to climate change. *Environmental Modeling and Assessment* 14(1): 1–16.

Kaika, M. (2005). *City of Flows: Modernity, Nature, and the City*. New York: Routledge.

Kasperson, R.E., Dow, K., Archer, E., Caceres, D., Downing, T., Elmqvist, T., Erikson, S., Folke, C., Guoyi, H., Vogel, C., Wilson, K., and Ziervogel, G. (2005). Vulnerable people and places. In: R. Hassan, R. Scholes and N. Ash (eds) *Ecosystems and Human Wellbeing: Current State and Trends* (Vol. 1, pp. 143–164). Washington, DC: Island Press.

Kates, R.W. (1985). The interaction of climate and society. In: R.W. Kates, J.H. Ausubel, and M. Berberian (eds) *Climate Impact Assessment: Studies of the Interactions of Climate and Society*. Chichester: Wiley.

Keil, R. and Desfor, G. (2004). *Nature and the City: Making Environmental Policy in Toronto and Los Angeles*. Tucson: University of Arizona Press.

Kelly, P.M. and Adger, W.N. (2000). Theory and practice in assessing vulnerability to climate change and facilitating adaptation. *Climatic Change* 47(4): 325–352.

Lewontin, R. and Levins, R. (2007). *Biology Under the Influence: Dialectical Essays on Ecology, Agriculture and Health*. New York: Monthly Review Press.

Liverman, D.M. (1990). Drought impacts in Mexico: climate, agriculture, technology, and land tenure in Sonora and Puebla. *Annals of the Association of American Geographers* 80(1): 49–72.

Massada, A.B., Radeloff, V.C., Stewart, S.I., and Hawbaker, T.J. (2009). Wildfire risk in the wildland–urban interface: a simulation study in Northwestern Wisconsin. *Forest Ecology and Management* 258(9): 1990–1999.

McCarthy, J., Canziani, O., Leary, N., Dokken, D., and White, K. (eds) (2001). *Climate Change 2001: Impacts, Adaptation and Vulnerability*. Contribution of Working Group II to the Third Assessment Report of the Intergovernmental Panel on Climate Change (IPCC). Cambridge: Cambridge University.

Moser, C. (1998). The asset vulnerability framework: reassessing urban poverty reduction strategies. *World Development* 26(1): 1–9.

Mustafa, D. (1998). Structural causes of vulnerability to flood hazard in Pakistan. *Economic Geography* 74(3): 289–305.

Mustafa, D. (2005). The production of urban hazardscape in Pakistan: modernity, vulnerability and the range of choice. *Annals of the Association of American Geographers* 95(3): 566–586.

O'Brien, K., Eriksen, S., Nygaard, L.P., and Schjolden, A. (2007). Why different interpretations of vulnerability matter in climate change discourses. *Climate Policy* 7: 73–88.

Orsi, J. (2004). *Hazardous Metropolis: Flooding and Urban Ecology in Los Angeles*. Berkeley: University of California Press.

Peet, R. and Watts, M. (2004). *Liberation Ecologies: Environment, Development, Social Movements* 2nd edition. New York: Routledge.

Pelling, M. (2003). Toward a political ecology of urban environmental risk. In: K.S. Zimmerer and T.J. Basset (eds) *Political Ecology: An Integrative Approach to Geography and Environment–Development Studies*. New York: Guildford Publications, 73–93.

Phillips, B., Thomas, D., Fothergill, A., and Blinn-Pike, L. (2009). *Social Vulnerability to Disasters*. London: Taylor & Francis.

Robbins, P. (2007). *Lawn People: How Grasses, Weeds, and Chemicals Make us Who we Are*. Philadelphia: Temple University Press.

Sanderson, D. (2000). Cities, disasters, and livelihoods. *Environment and Urbanization* 12(2): 93–102.

Swyngedouw, E. (2004). *Social Power and the Urbanization of Water*. Oxford: Oxford University Press.

Timmerman, P. (1981). Vulnerability, resilience and the collapse of society. *Environmental Monographs*. Toronto: No. 1 Institute for Environmental Studies, University of Toronto.

Turner, B.L.I., Kasperson, R.E., Matson, P.A., McCarthy, J.J., Corell, R.W., Christensen, L., Eckley, N., Kasperson, J.X., Luers, A., Martello, M.L., Polsky, C., Pulsipher, A., and Schiller, A. (2003). A framework for vulnerability analysis in sustainability science. *Proceedings, National Academy of Sciences of the United States of America* 100(14): 8074–8079.

Tyrell, J. (2010). National foreclosure rates rise to highest quarterly total we've ever seen. Retrieved from: <http://www.newjerseynewsroom.com/economy/national-foreclosure-rates-rise-to-highest-quarterly-total-weve-ever-seen>.

Wisner, B. (1993). Disaster vulnerability: scale, power, and daily life. *GeoJournal* 30(2): 127–140.

Wisner, B., Blaikie, P., Cannon, I., and Davis, B. (1994). *At Risk: Natural Hazards, People's Vulnerability and Disasters*. London: Routledge.

Wisner, B., Blaikie, P., Cannon, T., and Davis, I. (2004). *At Risk: Natural Hazards, People's Vulnerability, and Disasters* 2nd edition. London: Routledge.

Wolch, J., Pincetl, S., and Pulido, L. (2001). Urban nature and the nature of urbanism. In: M. Dear (ed.) *From Chicago to L.A.: Making Sense of Urban Theory*. London: Sage, 367–402.

Chapter 2

Development, Risk Momentum and the Ecology of Vulnerability: A Historical–relational Analysis of the 1991 Oakland Hills Firestorm

Gregory Simon

"Rockridge: a part of the city below yet apart from it."

Laymance Real Estate Company Brochure (1911)

The story of the 1991 Oakland Hills firestorm (Tunnel Fire) has been told in a variety of ways. As a story of broken lives and lost possessions for thousands of residents, as a display of bravery and heroism by community members and first responders, and as a reminder of the vulnerability of cities to destructive wildfires. These and other common narratives are accurate in emphasizing the magnitude and long lasting impact of the firestorm in the region's history. Indeed, over a 24-hour period the firestorm destroyed over 3000 homes, killed 25 people and seriously injured more than 150 others. Nearly 800 structures ignited in the first hour of the conflagration alone, and more than 300 per hour over the next seven hours (FEMA 1992).

Yet 20 years and many event narratives later, the story of the firestorm in relation to the region's historical development remains largely untold. This chapter emphasizes how the story of the firestorm *event* is part of a larger *process* of regional economic development and natural resource management. Using a historical–relational analytic approach, this chapter emphasizes how the production of vulnerability to the Tunnel Fire is deeply intertwined with a very specific set of regional development policies and land use practices oriented around wealth accumulation for property developers and home owners in the San Francisco Bay Area and around California. The story of the Oakland Hills demonstrates, first, how instrumentalist uses of tree cover – early timber extraction activities during the middle 1800s, and subsequent species introduction and afforestation strategies in the late 1800s and early 1900s – contributed to the production of vulnerable conditions in Oakland hillside areas. Second, this chapter reveals how conservative homeowner politics and State tax restructuring during the 1950s to

1970s further contributed to the generation of vulnerability in the city's hill and flatland regions.

Each story intertwines to emphasize the regional and historical context of vulnerability to the 1991 Tunnel Fire while also highlighting vulnerability as more than just a static condition or outcome. Rather, it is a condition of *effect* stemming from intentional regional development and planning policies, and also a state of *affect* that engenders further land use responses and produces new and enhanced levels of vulnerability. In this way, vulnerability is shown to accrue over time, build momentum and inscribe.

Vulnerability at the Wildland/Urban Interface (WUI): Moving Beyond the Snapshot Assessment

Scholarly attention to the study of vulnerability and WUI wildfires has focused primarily on defining and mapping the social and ecological attributes of risk and vulnerability, and developing snapshot risk assessments for human and non-human populations (Salas and Chuvieco 1994; DeBano et al. 1998; Haight et al. 2004; Bond and Keeley 2005; Hardy 2005; Galtie 2008; Massada et al. 2009; Keane et al. 2010). While this scholarship is immensely important, this chapter takes on a different, yet equally valuable, agenda by exploring vulnerability in its spatial–historical context. Rather than use static assessments of vulnerability to explain the WUI environment, this research suggests it is vulnerability itself that requires explanation.

Nearly 20 years after the Tunnel Fire, the conflagration has left a lasting legacy in the region for being the largest urban wildfire in terms of dwellings destroyed in California history. But the Tunnel Firestorm, like so many others, is also notable for being a deeply geographical event that is closely connected to historical policies, landscapes and vulnerabilities across space (Pyne 1997). Despite innumerable after-action-reports, retrospective editorials, and post-disaster assessments, important geographical questions about the fire and East Bay Hills still remain. Most notably, historical fire regime analysis of the hills region dating back to 1900 indicates that many neighborhoods are located in areas experiencing a high frequency of wildfires (see Figures 2.1 and 2.2). Given these apparent risks, this essay pivots around the following simple question: why were homes built into such a vulnerable landscape?

Expanding the Ontological Boundaries of Vulnerability: A Historical–relational Approach

This chapter suggests that addressing this important question requires retelling the story of the Tunnel Fire in the context of the city of Oakland, its environmental history, and its position within broader Bay Area and California growth economies.

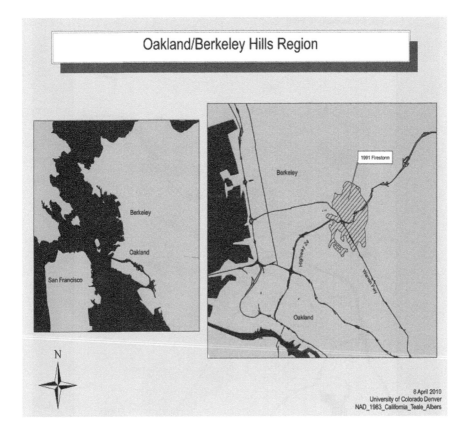

Figure 2.1   Firestorm area

As the fire historian Stephen Pyne remarks, "[fire] cannot be studied in itself; it is a profoundly interactive technology; it is what its context makes it" (2009, p. 446).

To expand our ontological commitment to the study of vulnerability this chapter demonstrates:

1. that vulnerability to WUI fires exists as a process that unfolds over time and across space;
2. that vulnerability can only be fully explained when evaluated in the context of broader economic and environmental histories which, in turn, highlights vulnerability as an effect of instrumentalist development policies; and
3. that vulnerability is never just produced as a mere planning ending or outcome; rather it is always in production, gaining momentum, at play, and affecting.

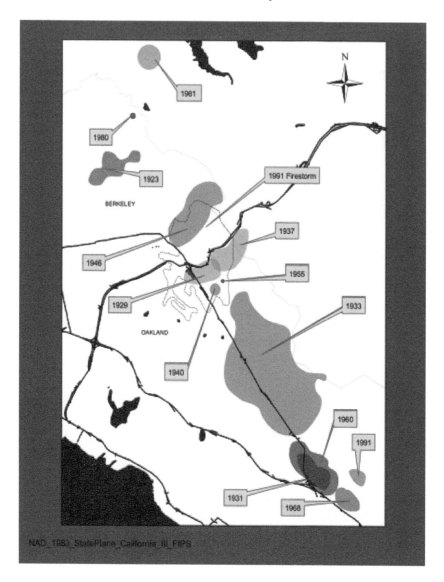

Figure 2.2   Historical fires in area

A historical–relational framework approach thus builds on snapshot accounts of vulnerability and towards a longitudinal analysis of the diverse social and ecological mechanisms influencing how vulnerability emerges, accumulates, shifts, and dissipates over time and space. This approach stands in contrast to more contemporary accounts that frequently isolate fire areas for analysis and obscure their connections across political and economic scales (Haight et al. 2005; Hardy 2005; Masada et al. 2009).

Considerable attention has been dedicated towards unearthing the social systemic dimensions of vulnerability, beginning with the work of Hewitt (1983), Wisner (1993), and Wisner et al. (2004) who all note the importance of highlighting the human aspects of so-called "natural" disasters. And yet, despite these advancements, dedicated analysis to the production of vulnerability is motivated by an uneven collection of data that, as Cutter et al. (2003) note, has produced considerably less information about the social aspects of vulnerability as compared to biophysical and built environment dimensions.

Expanding research on social vulnerability through a historical–relational framework requires utilizing key concepts and analytic approaches found in "pressure and release" and "human ecology" strands of vulnerability research. Following Pelling (2003), these frameworks of inquiry are useful for considering drivers of vulnerability (Liverman 1990; Mustafa 1998; Peet and Watts 2004) emanating from physical or biological conditions as well as "the cumulative progression of vulnerability, from root causes through to local geography and social differentiation" (Adger 2006, p. 272). This approach offers conceptual space for closely examining how political economies of resource use and normative planning and management interventions interact over time to influence the location and scope of vulnerability (Davis 1998; Orsi 2004; Wisner et al. 2004; Collins 2005, 2008, 2010; Mustafa 1998, 2005).

This framework, as applied to the Tunnel of Fire, also leverages theoretical insights and epistemological commitments within the field of urban political ecology (UPE) that seek to eschew modern binaries of human–nature and urban–rural, and highlight the mutually constitutive relationship between economies, policies and ecological systems across temporal and spatial scales (e.g., Cronon 1991; Gandy 2002; Swyngedouw 2004; Kaika 2005). This project therefore follows the recommendations of Zimmerer and Basset (2003) who note that, "diverse environmental processes interact with social processes, creating different scales of mutual relations that produce distinctive political ecologies" (p. 3). A historical–relational retelling of the Tunnel Fire also elucidates the deeply political and contentious process through which vulnerability to wildfires is produced, defined, and mitigated. These insights are informed by a second intellectual thrust within UPE, highlighting cities as complex ecological systems governed through power-laden politics and shaped by an unremitting commitment to economic development (Wolch et al. 2001; Keil and Desfor 2004; Dooling et al. 2006; Heynan et al. 2006; Robbins 2007; Brownlow 2008).

The Production of Vulnerability to WUI Fires: Conventional Approaches

While the vast majority of research on WUI wildfire vulnerability is directed towards snapshot assessments of vulnerability (see e.g., Salas and Chuvieco 1994;

DeBano et al. 1998; Haight et al. 2004; Bond and Keeley 2005; Hardy 2005; Galtie 2008; Massada et al. 2009; Keane et al. 2010), an emerging body of literature is beginning to assess the causes of vulnerability to WUI wildfires (and not just *that* certain populations are vulnerable). This area of research focuses specifically on identifying factors responsible for elevating levels of risk for human populations. These approaches fall within three overlapping sub-categories. A first sub-category of WUI wildfire research has focused on infrastructure and technology applications such as home construction architecture and roofing materials (Cohen 1999, 2000; Collins 2005), road width and accessibility for residents and emergency vehicles (Church and Sexton 2002; Cova and Johnson 2002; Cova 2005) and available water reservoir capacities (Xanthopoulos 2004). A second area of study focuses on the role of the physical environment in enhancing rates of vulnerability to wildfires at the WUI. Susceptibility to WUI fires, through fire intensity and rate of spread, is associated with climates containing extended periods of low humidity and prevailing dry winds (Westerling et al. 2003), topographic features that assist in the rapid movement of wildfires (Rehm and Mell 2009), and levels of maintained defensible space around homes to minimize the ignition of combustible vegetation cover (Gill and Stephens 2009). A third area of study focuses on fire management practices including insufficient attention to fire prevention in WUI areas (Schoenaegal et al. 2009) and the consequences of dedicated fire suppression in nearby wildland landscapes (Cohen 2008; Pyne 2008). This research area also calls into question coordination deficiencies between citizens, neighborhood organizations and local and regional fire prevention and mitigation agencies (Stephens et al. 2009).

The Tunnel Fire, as a subject of analysis, is indicative of firestorms in the US West, and has undergone close scrutiny within these three conventional forms of "vulnerability production" analysis (Rehm et al. 2001). Retrospective analysis of the 1991 Tunnel Fire has led fire experts and city agencies to scrutinize the built environment of the East Bay hills, citing high numbers of densely packed homes containing wood shingle sidings and shake roofs; and an extensive network of narrow, winding, and difficult to access roads (Oakland Police Department 1992; Alameda County Sheriff's Department 1992; FEMA 1992; City of Oakland 1992a). Similarly, the region's physical attributes came under immediate scrutiny, including the area's steep canyons, dense vegetation around homes, seasonal drought patterns, and dry Santa Ana Winds (City of Oakland 1992b; FEMA 1992; Oakland Police Department 1992). Inadequate levels of vegetation and fire-break management have also been well documented (City of Oakland 1992a). These retrospective approaches begin to reveal underlying factors causing vulnerability to wildfires at the WUI.

And yet, a historical–relational approach to vulnerability entails a much deeper engagement with political economic and multi-scale analysis. Building on a surprisingly small body of scholarship examining the historical, regional, and systemic dimensions of fire risk at the WUI (e.g., Davis 1998; Wolch et al. 2001; Collins 2005, 2008), this framework leverages an explicit acknowledgement of vulnerability's production to reassess the ontology of vulnerability as something more than a passive condition, outcome or social–ecological inscription. Instead this chapter articulates vulnerability as an active process, embedded within regional environmental and development histories, that *inscribes*.

Vulnerability Produced: Bay Area Development and Instrumentalist Timber Use

Logging, Road Infrastructure and Reverse Adaptation

This chapter's description of vulnerability and its production begins with large scale clear-cutting of redwoods trees (*Sequoia sempervirens*) occurring in the area as early as 1840. During this period "redwoods as tall as 300 feet and as wide as 32 feet" were captured and hauled to shipping points and sent for home and commercial construction purposes in the region during and immediately following the Gold Rush boom cycle (Bagwell 1982, p. 15). This lumber was also used to replace structures and resurrect portions of the burgeoning city of San Francisco that had burned during major fires in 1850 and 1851. In this way, development in San Francisco during the mid 1800s was aided to a great extent by a vast supply of east bay timber.

Increased commercial redwood lumber prices reflect this demand as prices skyrocketed from \$30/1000 board feet in 1847 to upwards of \$600 in 1849 (Bagwell 1982). By 1852, there were four steam sawmills operating in the Oakland Hills. The operation was so extensive that by 1860, hardly a tree remained (City of Oakland 1996). In lower hill areas, various farming activities including wheat and hay production ensued during subsequent decades. Many of these grains were used for feed within newly established cattle grazing outfits in the foothills region – an emerging set of land use practices that utilized the recently denuded landscape now free of coast live oak (*Quercus agrifolia*), redwoods and other native riparian woodlands.

The role of the east bay hills as a critical source of raw materials for the construction of San Francisco during its post-gold rush economic ascendance should not be underestimated. The Sierra Nevada foothills contained mineral deposits paving the way for wealth generation and frenzied infrastructure investments in the San Francisco bay area during the middle 1800s. Meanwhile, the east bay hills enclosed a ready supply of raw construction materials to facilitate actuating those investment objectives that, in turn, spurred a regime of continued material accumulation. For its ability to serve as a major source of timber, mineral and grazing resources, the East Bay Hills region stands as an early (and often overlooked) site of instrumentalist resource extraction – a role later played by the likes of California's Central Valley, Hetch Hetchy Valley, and North Coast forests.

As Gray Brechin notes in *Imperial San Francisco: Urban Power, Earthly Ruin*, when San Francisco grows, much like all great cities, "so does both its reach and its power to transform the nonhuman world on which its people depend". He continues, "there exists a critical *ecological* relationship between the city and the countryside, a relationship as applicable to modern San Francisco as to ancient Rome" (2006, p. xxix, italics in original). The historically resource rich Oakland Hills "countryside" played a crucial role in shaping, and indeed facilitating Oakland's development and economic relationship with San Francisco and other Bay Area cities.

A trajectory towards increased vulnerability in the Oakland Hills was set in motion as a result of this cross bay relationship. Logging, resource extraction, and land conversion activities during the mid-1800s not only altered the Oakland Hills landscape, they also introduced crude infrastructure and laid the foundations (and momentum) for subsequent housing development trajectories. Many current day roads that abut or cross through the Oakland Hills fire area (such as Redwood Road and Park Blvd) originally terminated at logging sites and were used to haul timber down slope to the Oakland Estuary and across the Bay to San Francisco. As Bagwell (1982) notes "several other East Bay roads (including present day Claremont and Thornhill Roads) also began as logging roads" (p. 18). While many of these roads have been significantly widened and modified, most remain in the same graded location as they existed during the mid to late 1800s.

Momentum towards increased home development, spurred by the introduction of hillside logging roads, occurred in two ways. On the one hand, road and public infrastructure in the area was financed primarily by the City of Oakland, and also by local banks who provided loans to public service corporations. For these entities, preexisting graded and reinforced thoroughfares were a cost efficient means of bringing roads to large real estate owners waiting to initiate home construction. On the other hand, these hill slope property owners did not sit idly by for infrastructure to reach them. Indeed, powerful landowners such as The Realty Syndicate were "not compelled to wait, as is the individual, upon the completion of corporation or municipal facilities," as they proceeded to "develop neighborhoods of a high class nature in absolutely new districts" (Blake 1911, p. 269). Such efforts by large real estate holders to improve their property through leveling and tree planting activities prior to the completion of adjacent road improvements, helped to attract public investments and "pull" housing developments (and paved roads) onto their property. Because of both cost efficient public expenditure imperatives and considerable upfront private investments alongside logging roads, these historical access routes became the pathway for new home developments in the North Oakland Hills.

This retelling of the Oakland Hills Fire thus constructively infuses the study of vulnerability with theories of "reverse adaptation" (Birkenholtz 2009; Simon 2010) by illustrating how urban planners and housing developers adapt to the form and location of assorted infrastructure such as utility roads – thereby granting infrastructural networks a certain level of agency to influence the environment, and levels of risk and vulnerability, around them. This perspective stands in contrast to a more widely-held belief that technology and infrastructure innovations adapt to human needs and planning objectives.

Afforestation, Eucalyptus, and the Real Estate Syndicate

The Claremont Hotel and Resort, which stands today as a landmark and visually dominating icon at the foot of the Oakland/Berkeley Hills, was first constructed alongside major resource extraction access roads in 1915. Built by a suite of local real estate developers, the hotel was erected in part to attract home construction investments and potential homebuyers to hillside areas located just south and upslope along these arterial routes. The hotel has thus come to represent the conjoining of two eras of regional development and wealth creation activities; first through its strategic placement at the base of former logging roads (present day Tunnel and Claremont Canyon roads), and second by serving as a hub connecting land development and construction activities to former timber extraction routes and the region's now defunct "Key Route" rail system. Indicative of the region's development along lines of estate-based wealth accumulation, the Claremont Hotel, Key Route system and ubiquitous Realty Syndicate land holdings in the Hills area – which all reinforced the other's profit-generating potential – all fell under the ownership and operation of two prominent developers: Frank Havens and Francis "Borax" Smith.

The Claremont Hotel signifies a transition in the Oakland Hills from productive logging area to a valued landscape falling under the speculative eye of housing real estate developers. While the introduction of crude road infrastructure helped generate access possibilities for developers, dedicated tree importation and reforestation efforts in subsequent decades further facilitated housing tract speculation, construction, and marketing efforts.

Early tree planting can be traced to the well-known essayist Joaquin Miller who, in 1886, purchased 69 acres and immediately planted pines, cypresses, acacia, and eucalyptus on his property (Nowak 1993). (See Figures 2.3 and 2.4 for images of the denuded landscape confronting Joaquin Miller and subsequent rapid afforestation around the Miller Homestead.) Forestry efforts from 1885–1893 were guided by the California State Forestry Board, which favored quick growing eucalyptus trees. Between 1910 and 1913, and nearly 50 years after the removal of virtually all tree cover in the hills region, Frank Havens planted approximately three million non-native eucalyptus (*Eucalyptus globulous*) and Monterey Pine (*Pinus radiata*) seedlings along the region's hills slopes. Eucalyptus trees were planted for commercial lumber speculation due to their purported lumber quality. During the next several decades, the fast growing eucalyptus indeed proved to be a hearty species tolerant of high winds, shallow soils and seasonal drought conditions (Tyrell 1999). Unfortunately for the timber industry, they also turned out to be a poor source of construction worthy lumber due to their interlaced wood fibers and irregular grain.

Despite having low construction utility, the Eucalyptus still served a crucial development function. Beginning in lower hills areas, these tree species were valued for their ability to beautify and increase land values in the hillside by replacing unsightly barren slopes with a bucolic residential environment suited for

Figure 2.3   Intensive logging during the mid-1800s left much of the
              Oakland Hills area void of tree cover. The Joaquin Miller
              homestead (1886) sits in the foreground

Figure 2.4   By 1913 the environment around the Miller homestead and
              much of the Oakland Hills changed considerably. Non-native
              and ornamental species can be seen around the property

affluent members of the San Francisco and East Bay business class (Nowak 1993). Following these successful Eucalyptus and Pine plantings, directors of the Realty Syndicate including Frank Havens and "Borax" Smith noted, "the increased value their holdings would have if plentifully timbered" (Oakland Tribune 1923). The Mahogany Eucalyptus and Land Company, which dominated property holdings in the Tunnel Fire area at the turn of the 19th century noted that, "This tree at this particular moment is in many instances the most valuable one on the face of the globe … The Company now sees plainly that it possesses a source of emolument higher than that of the average gold mine" (O'Brien 2005). By the early 1920s, this newly forested landscape containing vast Eucalyptus groves began to fulfill its prescribed development potential, as it quickly became populated by a number of housing subdivisions. The introduction of eucalyptus and other non-native tree species helped to deliver desired economic and ecological outcomes that, over time, produced regions notable for their elevated levels of fire vulnerability.

This historical narrative complies with recent calls to rematerialize geographical studies (Whatmore 2006) through the consideration of assemblage geographies (Robbins and Marks 2010) where "it is not only social actors working on nature, but bits of nature producing social action" (Shaw et al. 2010, p. 387). Eucalyptus are shown to be both agents of *effect* introduced to fulfill economic interests through the construction of *desirable* neighborhoods, and also agents of *affect* capable of being *desired* through a process that cultivates aesthetic and emotional connections for surrounding community members. This bi-modal condition is similar to road infrastructure originally constructed for its positive effect on resource extraction activities, while later becoming a technology of affect that attracted real estate developers to the area. It is through this landscape transition – from effectual to affectual – that vulnerability gains momentum, accumulates and actively inscribes onto the hillside.

Within its recent history, the North Hills region of Oakland "originated as a lumbering center" before transitioning into "a residential area best known for its spectacular views, forested character, winding streets, and hillside architecture"; a transformation marked by its eventual evolution into a "vacation retreat for San Franciscans" (City of Oakland 1998, p. 205). In 1923, e.g., the Oakland Hills witnessed a 900 percent increase in home construction over the previous five years (Bagwell 1982). If redwoods served as raw material for constructing a burgeoning San Francisco during the 1850s, Eucalyptus and Monterey Pine plantings during the 1920s served as the material basis for constructing a suburban respite from the hustle-bustle of San Francisco's now frenzied business environment. As one real estate booster put it, "these [San Francisco] new-comers … found themselves as far removed from the dirt and turmoil of the work-a-day world as if they had traveled fifty miles into the mountains" (Blake 1911, p. 261).

Utilizing the areas' new hillside tree cover and countryside image, Oakland Hills property owners leveraged the very wealth their hillside timber resources helped to create, to re-acquire capital investments from San Francisco and other Bay Area elites. As this chapter has shown, this transformation from a region exporting natural resource capital *to* the region, to an area actively receiving financial investments *from* the region has contributed to the production of vulnerability every step of the way.

Eucalyptus, Housing Stock, and Fire Vulnerability

The proliferation of eucalyptus trees for development speculation – introduced overtime as an important natural resource for both production and amenity oriented economies – has had the long-term effect of heightening fire risk and levels of social vulnerability in the area. According to a Federal Emergency Management Agency (FEMA) report issued after the Tunnel Fire, Eucalyptus and Monterrey Pine are "highly vulnerable to rapid fire spread" because they "release massive amounts of thermal energy when they burn. They also create flying brands, which are easily carried by the wind to start new spot fires ahead of the fire front" (FEMA 1992). As the historian Jared Farmer notes, Eucalyptus, "covered some 20 percent of the fire area and contributed an estimated 70 percent of the fuel load. The trees hardly caused the fire, but they did add to its intensity" (Farmer, forthcoming).

Communities ultimately seared by the Tunnel Fire became vulnerable over many decades as part of a process of vulnerability accumulation that directly benefited real estate owners and developers around the Bay Area. After the first commercial round of timber, conditions of vulnerability began to gain traction; slowly being inscribed into the hill slopes through road grading and rudimentary infrastructure construction. Fifty years later, home developments were constructed along these and other access corridors in areas that, as Figure 2.4 depicts, contain historically high levels of susceptibility to wildfires. Vulnerability within these communities was further augmented through the introduction and maturation of a property value-enhancing replacement tree cover comprised, to a large extent, by flammable Eucalyptus and Monterrey Pine species. As the 1986 Oakland Hill Area Specific Plan mentions, "In addition to naturally-occurring fires, the potential for accidental fires has increased as a result of urbanization in the area. With increasing urbanization, plant species such as eucalyptus and highly flammable ornamental vegetation have been introduced to the area" (p. 112). Of course, the houses themselves added to net vulnerability by increasing the likelihood and severity of fires in the area and adding substantially to the regions fuel load. According to the City's safety plan, "Most of the severity of Oakland's fire hazard stems from the presence of residential neighborhoods and ... the many wooden structures in the area".

As Adger (2006) aptly notes, vulnerability "does not exist in isolation from the wider political economy of resource use" (p. 270). Indeed, deforestation and afforestation trends in the Oakland Hills were instrumental in growing the Bay Area economy, as capital flowing out from Oakland eventually turned course, delivering substantial financial payments through housing and land speculation channels back across the San Francisco Bay into Oakland. Vulnerability to the Oakland Hills Fire, and other historical East Bay fire events, can thus be productively viewed as the byproduct of deeply intertwined regional economic and environmental histories.

Vulnerability Produced: California Tax Revolt, Depleted Budgets and Uneven Risk Offsetting

Proposition 13 and Diminished City Revenues

In 1978, the State of California witnessed a dramatic change in its taxation and state revenue structure. Property taxes directly linked to quickly rising inflation rates and skyrocketing property values began to jeopardize the preexisting tax system. Members of an emerging conservative anti-tax (or "tax revolt") movement began agitating to overturn the perceived burdensome tax structure. They launched a massive publicity campaign attacking rising property tax payments, and criticizing revenue dispersements for public expenditures in older city segments far removed from self-described "overtaxed" suburban homeowners.

Public sentiment across California was put to the vote in 1978 and the tax revolt initiative won overwhelmingly 64.8 percent to 35.2 percent, leading to the passage of Proposition 13 – the nation's first comprehensive tax limitation measure. Voting was not equal across all cities however. In Alameda County, Oakland suburbs like San Leandro, Hayward and Fremont approved the measure by more than 70 percent; while the residents of Oakland narrowly rejected the measure by 52 percent. This voting pattern underscores a general bifurcation in post-war metropolitan development around the state and nation: a substantial redistribution of investments, tax revenues and wealth into suburban areas and away from core cities like Oakland, which in turn contributed to the slow decay and underdevelopment of these older metropolitan areas.

The effects of Proposition 13 were nothing short of profound for California's economy and intra-state circulation of financial capital. They set maximum property tax rates at 1 percent of the total value of properties, and restricted maximum increases in assessed value of property to a mere 2 percent from one year to the next. Of even greater effect, its passage mandated that property could only be revalued under a transfer of ownership. The impact was particularly dramatic for cities like Oakland with aging infrastructure and a large working class population. Cities like Oakland have historically relied on local property tax revenues to support sizeable public works programs.

At the time of Proposition 13's passage, 90 percent of Fire Department budgets in California were funded through local property tax revenues (Brownlee in Self 2003). Across California more generally, counties drew 33 percent of their revenue from property taxes prior to Proposition 13, while by 1996 that number dropped to 12 percent for counties around the state. At the city level, the proportion of revenues drawn from property taxes decreased from 16 percent to 8 percent over the same period (Chapman 1998).

Reduced Fire Department Budgets and Vulnerability Intensified

Urban residents in cities like Oakland are now paying for many services through price systems composed of fees and charges rather than general citywide revenue streams such as the property tax. And yet, despite these new and augmented revenue sources, funding impacts on the Oakland Fire Department have been unmistakable. The passage of Proposition 13 increased vulnerability throughout the city by sharply reducing city tax revenues, resulting in underfunded fire response, mitigation and retrofitting programs (Self 2003). According to FEMA, the Oakland hills and the coastal region more generally were "particularly vulnerable in the fall of 1991, after five years of drought, several months with no recorded precipitation, and reduced efforts to control wildland interface fires due to State and local budget limitations" (1992, p. 2). The report went on to note, "Before budget reductions in the 1970s and 80s, [The Oakland Fire Department] was recognized as one of the strongest fire suppression departments in the western United States. The budget limitations reduced the number of companies in service and the staffing on each company. Several stations were closed during this period" (FEMA 1992, p. 50).

Budget shortfalls in Oakland helped engender conditions characterized by increased fire risk. For example, it is well documented in the City of Oakland and elsewhere that damaged and partially dead trees pose a significant fire risk. A deep freeze in 1972 and again in 1990 damaged numerous Eucalyptus groves and contributed to the development of highly flammable vegetation cover. Yet, according to the Oakland General Plan, budget reductions led to conditions where "the City lacks the funds to completely restore its damaged or dead vegetation" (City of Oakland 1996, pp. 3–38).

Moreover, according to City officials, Oakland's Fire Department operational units were overextended in 1991, leading to diminished emergency response and mitigation capabilities. As a result, the Office of the Fire Chief suggested that a concerted effort to "reinstate Battalion 5, in the north Oakland hill area, will be requested in the Budget for Fiscal Year 1993–1994". According to the Fire Chief, extra financial resources would be demanded to reinstate this position, as its absence due to budget restrictions, was of great detriment to the overall 1991 fire response. Doing so "will decrease our command span of control from ten (10) to one (1), to approximately 7.5 to one. The average span of control ratio for Oakland's population and square miles is approximately six to one" (City of Oakland 1992a).

In light of budgetary constraints, many fire departments now rely on a regional network of response agencies. This model of flexible and shared emergency response governance is exceedingly more cost efficient for participating municipalities than fully staffed, autonomous units. A fire disaster the size of the Tunnel Fire, in particular, requires the involvement of multiple agencies and complex interagency coordination. As the Alameda County After Action Report notes, " the magnitude of the firestorm's devastation taxed the city far beyond its ability to respond without assistance" (p. 20). And yet, interagency communication breakdowns and technological incompatibilities rendered the hillside community under-protected at a time when they were particularly reliant on external agency assistance. The tunnel fire illustrates how residents in the City of Oakland – who had become increasingly dependent on "flexible" and "efficient" cross-jurisdictional support to minimize levels of vulnerability – were left abandoned by this interagency response structure. For example, according to Tom Bates, California Assemblyman at the time of the Tunnel fire, "nozzle hook-ups for Oakland ... actually had a smaller size than the other districts. So, when firefighters came from areas, they could not plug into the Oakland hydrants." Mr Bates continued, "Communications broke down. They could not communicate with one another adequately because the radios were on different wave lengths". Different counties and city fire departments in the region "had mutual fire agreements ... and what they found was that when people arrived, they did not have a very good command structure ... It was pretty much chaos" (State of California 2001) Along with these technical and communication incompatibilities, many of these assisting agencies themselves were undergoing similar constraints as a result of severe budget cuts (Pincetl et al. 2008).

Overall, the "efficient" fire response and mitigation model that emerged post-budget reductions was itself rendered vulnerable to technical and communication limitations – thus making the system inadequate in its response capacity. The Tunnel Fire should therefore serve as a reminder of the potential pitfalls associated with heavy reliance on regional fire response frameworks implemented to cope with budget constraints.

Vulnerability Through City Revenue Offsetting and Home Construction: Postwar Suburban Roots

Tax restructuring has facilitated the pursuit of new developments in the East Bay Hills area. While big box retailers and car dealerships are among the largest tax revenue generators for cities and counties, home developments and subdivisions can also generate extra revenues, particularly in areas like the North Oakland Hills that are unsuitable for most large commercial developments, and that contain high property values and tax revenue potential (Chapman 1998). Accordingly the *1986 North Oakland Hill Area Specific Plan* states, "given the assumed value of new homes in the North Oakland Hills Area, and the significant level of property taxes generated, the net fiscal impact of development on public services is positive ...".

The report goes on to note, however, that this net fiscal benefit to the city of Oakland holds true "... unless an attempt is made to operate a new fire station" (p. 122). Here, City officials describe how tax revenues from new developments in the Oakland Hills can have a positive fiscal impact on the city and offset decreased property tax revenues, such as from the passage of Proposition 13. But net revenues will occur only without the construction and maintenance of a new fire station. This means that populating the hill slope in a manner that makes up for lost revenues, requires not only intentionally placing homes within a landscape historically vulnerable to frequent wildfires, it requires doing so without additional fire protection. As one city report soberly put it, "New residential development will significantly increase the potential for loss of life and damage to property from fire hazards in the North Oakland Hills, especially given the poor accessibility" (City of Oakland 1986). The ripple effect of lost revenues in cities like Oakland has therefore had the consequence of putting more households in harms way, while simultaneously underfunding fire response agencies and denying new developments adequate fire protection – this despite the Fire Agency's plea for increased communication center capacity, and the addition of extra fire battalion units in the hills area (City of Oakland 1992a).

Looking back further into the region's social and economic history, it is evident that increased home construction in the Oakland Hills in pursuit of greater city tax revenues holds its roots in the post-war suburbanization of the greater east bay. During this period, numerous suburban communities received massive amounts of public and private investments to attract homeowners, industry, and financial capital – marking a redistribution of wealth, property values, and tax revenues that directly undercut the economic best interests of core cities like Oakland. This process was spurred by large government subsidies in the suburban housing market, a collection of construction entrepreneurs and city boosters, and a diverse population of white homeowners seeking financial independence and low property taxes. Through market forces and federal urban policy retrenchment, the suburban landscape was converted into various forms of capital: increased property values for homeowners, direct profits for developers and city boosters, and taxes for public agencies (Self 2003). The results of regional suburbanization became apparent in the rapid development of suburban landscapes and the underdevelopment of Oakland during the post-war period.

And so, prior to the passage of Proposition 13, Oakland was already actively attempting to offset these lost revenues by constructing new housing units in hillside areas. From the late 1950s to the mid 1970s city builders in Oakland erected hundreds of new high-density units in the hill areas immediately adjacent to the Caldecott Tunnel – the very Tunnel area that sparked the firestorm (and its name). Unsurprisingly, units in the Parkwood Apartments and Hiller Highlands complex were the first units to go up in flames during the 1991 conflagration.

This post-war era of suburbanization dovetails with the 1970s tax revolt policy era in two important ways: first, each had the intent and effect of protecting the individual financial interests of wealthy white suburbanites while reordering investment and tax revenue streams away from older, working class cities with a higher percentage of minority residents; and second, vulnerability to wildfires increased through patterns of revenue impingements affecting tax dependent city programs, and as a result of the increased need for public officials to generate more tax revenues from high fire risk areas within the city's boundaries.

The intended effect of Proposition 13 and the conservative homeowner movement more generally was to limit property taxation and thus promote the consolidation of wealth amongst home and private business owners. These broad level monetary distribution adjustments went on to affect city fire prevention and response capabilities as a result of budgetary constraints while also presenting an economic rationality for Oakland to permit the construction of more homes in high risk fire areas. Once again, the production of vulnerability is shown to proceed through a step-wise process where local levels of vulnerability gain momentum and grow over time as a consequence of regional economic transformations.

When described in its historical–regional context then, the Tunnel fire presents a useful case for constructively building on the work of Pincetl et al. (2008), who detail how tax restructuring and revenue reapportioning across space can deplete funding for crucial city fire prevention activities; and also the work of Collins (2005 and 2008), who describes how development speculation for housing communities increase patterns of vulnerability. Bringing these insights together – vulnerability through city revenues impingements and amenity-based housing development – this essay advances theories on the deeply relational nature of vulnerability's production as it grows and accumulates within the context of California's emerging ethos of liberal economic ideology during the 1970s.

California Conservative Homeowner Politics and the Uneven Production of Local Vulnerability

As Robert Self (2003) has noted in his groundbreaking essay "American Babylon: Race and Struggle for Postwar Oakland", Proposition 13 marked a seminal moment within California's broad cultural and economic ideological shift towards the pursuit and maintenance of neopopulist conservative homeowner policies predicated on individual rights, estate-based wealth protection and a nearsighted commitment to social responsibility. In the face of a suburban growth politics – which had successfully grown over the previous three decades culminating in the overthrow of conventional structures of taxation under Proposition 13 – metropolitan core areas like Oakland experienced tax revenue losses, leading to patterns of government retrenchment and underfunded fire departments.

As Proposition 13 voting patterns indicate, this was a movement largely driven by white suburban and rural homeowners. However, as one digs deeper into the spatial pattern of Proposition 13 voting in Oakland, an interesting trend emerges. Voters in the Oakland hills tended to follow suburban tendencies by voting in favor of the proposition by nearly two to one, while downtown Oakland and the flatlands, comprised much more heavily by working class minorities, voted two to one *against* the measure (Kemp 1980). This intra-city division in voting underscores the notion of Oakland as a "tale of two cities", where residents of the Oakland hills and flatlands areas clearly held different views on what constitutes acceptable property tax rates and levels of revenue sharing. Indeed, the quote opening this essay can be productively evoked, as the hill residents largely perceived of themselves as "a part of the city" yet fiscally "apart from it".

As Mike Davis (1998) has noted, there exists a multi-level system of fire risk subsidization to buoy the widely held acceptance of (and contribution to) risk by hillside homeowners. This system of vulnerability-offsetting is supported to a large extent by the insurance industry. If residents can afford insurance rates and fees, they can effectively pay for the right to live in highly vulnerable areas. This is a privilege many others in working-class cities like Oakland cannot afford. Elevated fire risk, in terms of potential for lost property, possession and/or life, is therefore a burden that is unevenly felt by impoverished members of society who cannot afford high insurance premiums (or who live in tenements that do offer such coverage). The dire social consequences of underfunded and understaffed fire departments thus prove to be highly uneven across race and class based categories. As Robert Self (2003) opines, the story of Proposition 13 is as much about the suburbs as it is about life and livelihoods within inner cities.

There also exists a tiered structure of post-disaster redevelopment policies facilitating risk reduction. For example, immediately after the firestorm, the State of California paid an estimated $15 million to local governments in the form of public disaster assistance. This included payments directly to local governments, loans to owner-occupied and rental properties, individual and family grants and homeowner property tax deferrals. Meanwhile federal grants were issued for an estimated $42 million to state and local governments to recover these and other incurred costs (State of California 1991).

The influence of these pre- and post-disaster risk offsetting policies should not be trivialized, as they help Oakland Hills residents and members of other suburban communities rationalize their decision to abet chronic disinvestments in citywide fire response and prevention agencies. After all, as Rodrigue (1993) notes, although hillside communities live in fire-prone areas, they do not hold a similar level of vulnerability to wildfires as less privileged residents in flatland areas with minimal access to vulnerability-offsetting resources. Post-war conservative homeowner politics, when coupled with private and government risk reduction measures, has had the effect as Swyngedouw (2000, p. 53) suggests, of shielding "the bodies of the powerful while leaving the bodies of the poor to their own devices".

The two stories within this chapter culminate then, after nearly 150 years of vulnerability-in-production, with the *fullest* burden of risk to urban fires besetting flatland community members largely detached from the region's history of wealth accumulation and instrumentalist land use policy.

Conclusion

This chapter has endeavored to show how the concept of vulnerability can be productively leveraged as a starting point for analysis, and not merely as an outcome to be analyzed. Conventional descriptions of WUI wildfire disasters like the Tunnel Fire tend to focus on social/ecological conditions and attributes found within the spatial boundaries of fire events, while largely ignoring historical and multi-scale factors that create conditions of vulnerability. This chapter advocates for a historical–relational framework that illuminates local land use changes and levels of vulnerability, while simultaneously highlighting the historical machinations of regional political economies of development, instrumentalist natural resource policies, and conservative social movements. The preceding pages have examined early resource extraction and property speculation activities closely linked to the development of nascent Bay Area townships, and also the implications of an emerging suburban homeowner politics and subsequent statewide tax restructuring policies. While these two narratives by no means represent the totality of factors influencing levels of risk, significant new story lines materialize which advance our understanding of factors that facilitate vulnerability's production in Oakland and other cities around the Western United States. Meanwhile, the use of vulnerability as an organizing concept has revealed new insights that connect the Tunnel Fire event to the city of Oakland, its environmental history, and its position within broader Bay Area and California growth economies. Interestingly, nearly 150 years of produced vulnerability has created a landscape of risk to urban fires that most heavily burdens community members largely detached from the region's history of economic development, wealth accumulation and instrumentalist land use planning policy.

As a result of this analysis, conceptual advancements of vulnerability as an active process begin to emerge. As this chapter has suggested, vulnerability accumulates and gains momentum over time – as a condition of effect stemming from regional economic development and profit seeking land use planning policies, and also as a state of affect that engenders further land use policy responses and produces new and enhanced levels of vulnerability. Vulnerability is thus never static – it is instead vicissitudinous, mutable and spatially uneven. Nor is vulnerability dormant – it is instead active, always in production, and affecting. Vulnerability is both inscribed and continually inscribing. Importantly, insights from this chapter can, in turn, generate crucial lessons for planning agencies and residents in other fire prone areas around the world by describing the acute and enduring social vulnerability implications of apparently "benign" contemporary land use planning and natural resource management decisions.

References

Adger N. 2006. Vulnerability. *Global Environmental Change* 16:3, 268–281.

Alameda County Sherrif's Department. 1992. The 1991 East Bay Hills firestorm: after-action report. San Leandro: Office of Emergency Services

Bagwell B. 1982. *Oakland: The Story of a City*. Novato: Presidio Press

Birkenholtz T. 2009. Irrigated landscapes, produced scarcity, and adaptive social institutions in Rajasthan, India. *Annals of the Association of American Geographers* 99:1, 118–137.

Blaikie P. and Brookfield H. 1987. *Land Degradation and Society*. London: Routledge.

Blake E. (ed.) 1911. *Greater Oakland*. Oakland: Pacific Publishing

Bond W.J. and Keeley J.E. 2005. Fire as a global "herbivore": the ecology and evolution of flammable ecosystems. *Trends in Ecology & Evolution* 20, 387–394.

Braun B. 2005. Writing a more-than-human urban geography. *Progress in Human Geography* 29, 635–650.

Brechin G. 2006. *Imperial San Francisco: Urban Power, Earthly Ruin.* Berkeley: University of California Press.

Brownlee. California taxes: historical roots and the property tax: will it survive? Brief on Proposition 13 prepared by Assembly Committees on Local Government and Revenue and Taxation, IGS.

Brownlow A. 2008. *A Political Ecology of Neglect: Race, Gender, and Environmental Change in Philadelphia.* Saarbrucken: Verlag DM.

California State Board of Equalization. 2010. California City and count sales and use tax rates. 71. October 1, 2010.

Chapman J.I. 1998. Proposition 13: some unintended consequences. Public Policy Institute of California.

Church R. and Sexton R. 2002. Modeling small area evacuation: an existing transportation infrastructure impede public safety? Final report, Caltrans Testbed Center for Interoperability Task Order 3021.

City of Oakland. 1986. North Oakland Hill area specific plan. The Oakland City Planning Department.

City of Oakland. 1992a. The Oakland Fire Department's response to the Office of Emergency Service report: The East Bay Hills fire or October 20, 1991. Office of the Fire Chief.

City of Oakland. 1992b. The Oakland Hills firestorm: after-action report. Oakland: Office of the City Manager.

City of Oakland. 1996. Open Space Conservation and Recreation (OSCAR). An element of the general plan.

City of Oakland. 1998. Envision Oakland. An element of the general plan.

Cohen J.D. 1999. Reducing the wildland fire threat to homes: where and how much? USDA Forest Service general technical report PSW-GTR-173.

Cohen J.D. 2000. Preventing disaster: home ignitability in the wildland–urban interface. *Journal of Forestry* 98, 15–21.

Cohen J. 2008. The wildland urban interface fire problem: a consequence of the fire exclusion paradigm. *Forest History Today: A Publication of the Forest History Society* Fall, 20–26.

Collins T. 2005. Households, forests, and fire hazard vulnerability in the American West: a case study of a California community. *Global Environmental Change B: Environmental Hazards* 6:1, 23–37.

Collins T. 2008. The political ecology of hazard vulnerability: marginalization, facilitation and the production of differential risk to urban wildfires in Arizona's White Mountains. *Journal of Political Ecology* 15, 21–43.

Collins T. 2010. Marginalization, facilitation, and the production of unequal risk: the 2006 Paso del Norte Floods. *Antipode* 42:2, 258–288.

Cova T.J. and Johnson J.P. 2002. Microsimulation of neighborhood evacuations in the urban–wildland interface. *Environment and Planning A* 34:12, 2211–2229.

Cova T.J. 2005. Public safety in the urban–wildland interface: should fire-prone communities have a maximum occupancy? *Natural Hazards Review* 6:3, 99–108.

Cronon W. 1991. *Nature's Metropolis: Chicago and the Great West*. New York: W.W. Norton.

Cutter S.L., Mitchell J.T. and Scott M.S. 2000. Revealing the vulnerability of people and places: a case study of Georgetown County, South Carolina. *Annals of the Association of American Geographers* 90:4, 713–737.

Davis M. 1998. *Ecology of Fear: Los Angeles and the Imagination of Disaster*. New York: Metropolitan Books.

DeBano L.F., Neary D.G. and Folliott P.F. 1998. *Fire's Effects on Ecosystems*. New York: John Wiley & Sons.

Dooling S., Simon G. and Yocom K. 2006. Place-based urban ecology: a century of park planning in Seattle. *Urban Ecosystems* 9:4, 299–321.

Eakin H. and Luers A. 2006. Assessing the vulnerability of social-environmental systems. *Annual Review of Environment and Resources* 31, 365–394.

Farmer L. In Preparation. *Trees in Paradise: A California History*.

Federal Emergency Management Agency (FEMA). 1992. The East Bay Hills Fire Oakland–Berkeley, California. U.S. Fire Administration/technical report series USFA-TR-060.

Galtie J.F. 2008. GIS-supported modeling and diagnosis of fire risk at the wildland urban interface. A methodological approach for operation management. In: *Modelling Environmental Dynamics: Advances in Geomatic Solution* (eds) M. Paegelow and M.T.C. Olmedo. Springer.

Gandy M. 2002. *Concrete and Clay: Reworking Nature in New York City*. Cambridge, MA: MIT Press.

Gill A.M. and Stephens S.L. 2009. Scientific and social challenges for the management of fire-prone wildland–urban interfaces. *Environmental Research Letters* 4 034014.

Gregory I.N. and Healey R.G. 2007. Historical GIS: structuring, mapping and analysing geographies of the past. *Progress in Human Geography* 31, 638–653.

Haight R.G., Cleland D.T., Hammer R.B., Radeloff V.C. and Rupp T.S. 2004. Assessing fire risk in the wildland-urban interface. *Journal of Forestry* 102:7, 41–48.

Hardy C.C. 2005. Wildland fire hazard and risk: problems, definitions, and context. *Forest Ecology and Management* 211, 73–82.

Harvey D. 1996. *Justice, Nature and the Geography of Difference*. Oxford: Blackwell Publishers.

Hewitt K. 1997. *Regions of Risk. A Geographical Introduction to Disasters*. Essex: Addison Wesley Longman.

Heynen N., Kaika M. and Swyngedouw E. 2006. *In the Nature of Cities: Urban Political Ecology and the Politics of Urban Metabolism*. New York: Routledge.

Hoffer P. 2006. *Seven Fires: The Urban Infernos that Reshaped America*. New York: Public Affairs.

Ionescu C., Klein R.J.T., Hinkel J., Kumar K.S.K. and Klein R. 2009. Towards a formal framework of vulnerability to climate change. *Environmental Modeling and Assessment* 14:1, 1–16.

Kaika M. 2005. *City of Flows: Modernity, Nature, and the City*. New York: Routledge.

Kasperson R.E., Dow K., Archer E., Caceres D., Downing T., Elmqvist T., et al. 2005. Vulnerable people and places. In: R. Hassan, R. Scholes and N. Ash (eds) *Ecosystems and Human Wellbeing: Current State and Trends* (Vol. 1, pp. 143–164). Washington, DC: Island Press.

Keane R.E., Drury S.A., Karau E.C., Hessburg P.F. and Reynolds K.M. 2010. A method for mapping fire hazard and risk across multiple scales and its application in fire management. *Ecological Modeling* 221:1, 2–18.

Keil R. and Desfor G. 2004. *Nature and the City: Making Environmental Policy in Toronto and Los Angeles*. Tucson, AZ: University of Arizona Press.

Kelly P.M. and Adger W.N. 2000. Theory and practice in assessing vulnerability to climate change and facilitating adaptation. *Climatic Change* 47:4, 325–352.

Kemp R.L. 1980. *Coping with Proposition 13*. Farnborough: Lexington Books

Liverman D.M. 1990. Drought impacts in Mexico: climate, agriculture, technology, and land tenure in Sonora and Puebla. *Annals of the Association of American Geographers* 80:1, 49–72.

Massada A.B., Radeloff V.C., Stewart S.I. and Hawbaker T.J. 2009. Wildfire risk in the wildland–urban interface: a simulation study in Northwestern Wisconsin. *Forest Ecology and Management* 258:9, 1990–1999.

McCarthy J., Canziani O., Leary N., Dokken D. and White K. (eds) 2001. *Climate Change 2001: Impacts, Adaptation and Vulnerability*. Contribution of working group II to the third assessment report of the Intergovernmental Panel on Climate Change (IPCC). Cambridge: Cambridge University.

Mustafa D. 1998. Structural causes of vulnerability to flood hazard in Pakistan. *Economic Geography* 74:3, 289–305.

Mustafa D. 2005. The production of urban hazardscape in Pakistan: modernity, vulnerability and the range of choice. *Annals of the Association of American Geographers* 95:3, 566–586.

Neumann R.P. 2010. Political ecology II: theorizing region. *Progress in Human Geography* 34, 368–374.

Nowak D.J. 1993. Historical vegetation change in Oakland and its implication for urban forest management. *Journal of Arboriculture* 19:5.

NRDC. 2008. Hotter and drier: the west's changed climate. Report by the Natural Resources Defense Council and the Rocky Mountain Climate Organization. New York: NRDC Publications.

Oakland Police Department. 1992. Hill fire disaster: after action report. Oakland, CA.

O'Brien K., Eriksen S., Nygaard L.P. and Schjolden A. 2007. Why different interpretations of vulnerability matter in climate change discourses. *Climate Policy* 7, 73–88.

Office of Emergency Services. 1992. The East Bay Hills fire – a multi-agency review of the October 1991 fire in the Oakland/Berkeley Hills. Sacramento: East Bay Hills Fire Operations Review Group, Governor's Office.

Orsi J. 2004. *Hazardous Metropolis: Flooding and Urban Ecology in Los Angeles.* Berkeley, CA: University of California Press.

O'Sullivan D. 2005. Geographical information science: time changes everything. *Progress in Human Geography* 29:6, 749–56.

Peet R. and Watts M. 2004. *Liberation Ecologies: Environment, Development, Social Movements* 2nd Edition. New York: Routledge.

Pelling M. 2003. Toward a political ecology of urban environmental risk. In: K.S. Zimmerer and T.J. Basset (eds) *Political Ecology: An Integrative Approach to Geography and Environment–Development Studies.* New York: Guildford Publications, 73–93.

Pincetl S., Rundel P.W., De Blasio J.C., Silver D., Scott T., Keeley J.E. and Halsey R. 2008. It's the land use not the fuels: fires and land development in Southern California. *Real Estate Review* 37, 25–42.

Pyne S.J. 1997. *Fire in America: A Cultural History of Wildland and Rural Fire.* Seattle: University of Washington Press.

Pyne S.J. 2008. Spark and sprawl: a world tour. *Forest History Today: A Publication of the Forest History Society* 4–10.

Pyne S.J. 2009. The human geography of fire: a research agenda. *Progress in Human Geography.*

Radeloff V.C., Hammer R.B., Stewart S.I., Fried J.S., Holcomb S.S. and McKeefry J.F. 2005. The wildland–urban interface in the United States. *Ecological Applications* 15:3, 799–805.

Radke J. 1995. Modeling urban/wildland interface fire hazards within a geographic information system. *Geographic Information Science* 1:1, 7–20.

Rehm R.G., Hamins A., Baum H.R., McGrattan K.B. and Evans D.D. 2001. Community-scale fire spread. In: K.S. Blonksi, M.E. Morales and T.J. Morales (eds) *Proceedings of the California 2001 Wildfire Conference: 10 Years After the 1991 East Bay Hills Fire, 10–12 October 2001,* Oakland, CA: University of California, Forest Products Laboratory, Technical Report 35.01.462. (pp. 126–139). Richmond, CA.

Rehm R.G. and Mell W. 2009. A simple model for wind effects of burning structures and topography on wildland–urban interface surface-fire propagation. *International Journal of Wildland Fire* 18:3, 290–301.

Robbins P. 2007. *Lawn People: How Grasses, Weeds, and Chemicals Make us Who we Are.* Philadelphia: Temple University Press.

Robbins P. and Marks B. 2010. Assemblage geographies. In: *The Sage Handbook of Social Geographies* (ed.) S. Smith, R. Pain, S.A. Marston and J.P. Jones III, 176–94. Beverly Hills, CA: Sage.

Rocheleau D. 2008. Political ecology in the key of policy: from chains of explanation to webs of relation. *Geoforum* 39:2, 716–727.

Rodrigue C.M. 1993. Home with a view: chaparral fire hazard and the social geographies of risk and vulnerability. *The California Geographer* 33, 105–118.

Salas F.J. and Chuvieco E. 1994. Geographical information systems for wildfire risk mapping. *Wildfire* 3:2, 7–13.

Sayre N. 2005. Ecological and geographical scale: parallels and potential for integration. *Progress in Human Geography* 29, 276–290.

Schoennagel T., Nelson C.R., Theobald D.M., Carnwath G.C. and Chapman T.B. 2009. Implementation of national fire plan treatments near the wildland–urban interface in the western United States. *2009 proceedings of the National Academy of Sciences of the United States.*

Self R.O. 2003. *American Babylon: Race and the Struggle for Postwar Oakland.* Princeton: Princeton University Press.

Shaw I.G.R., Robbins P.F. and Jones J.P. III. 2010. A bug's life and the spatial ontologies of mosquito management. *Annals of the Association of American Geographers* 100, 373–392.

Simon G. 2010. Mobilizing cookstoves for development: a dual adoption framework analysis of collaborative technology innovations in Western India. *Environment and Planning A* 42, 2011–2030.

State of California. 1991. The 1991 East Bay firestorm. California Legislative State Senate Torres, R., Chairman. Senate Committee on Toxics and Public Safety Management.

State of California. 2001. Oral history interview with Tom Bates. California State Archives State Government Oral History Program. By McGarrigle, L. Regional Oral History Office, University of California.

Stephens S.L., Adams M., Hadmer J., Kearns F., Leicester B., Leonard J. and Moritz M. 2009. Urban–wildland fires: how California and other regions of the US can learn from Australia. *Environmental Research Letters* 4 014010

Swyngedouw E. 2000. The Marxian alternative: historical geographical materialism and the political economy of capitalism. In: T. Barnes and E. Sheppard (eds) *Reader in Economic Geography.* Oxford: Blackwell, 41–59

Swyngewouw E. 2004. *Social Power and the Urbanization of Water.* Oxford: Oxford University Press.

Theobald D.M. and Romme W.H. 2007. Expansion of the US wildland–urban interface. *Landscape and Urban Planning* 83, 340–354.

Turner B.L.I., Kasperson R.E., Matson P.A., McCarthy J.J., Corell R.W., Christensen L., Eckley N., Kasperson J.X., Luers A., Martello M.L., Polsky C., Pulsipher A. and Schiller A. 2003. A framework for vulnerability analysis in sustainability science. *Proceedings, National Academy of Sciences of the United States of America* 100:14, 8074–8079.

Tyrell I. 1999. *True Garden of the Gods: California–Australian Environmental Reform `860-1930.* Berkeley: University of California Press

Walker R. 2007. *The County in the City: The Greening of the San Francisco Bay Area*. Seattle: University of Washington Press.

Westerling A.L., Brown T.J., Gershunov A., Cayan D.R. and Dettinger M.D. 2003. Climate and wildfire in the western United States. *Bulletin of the American Meteorological Society* 84:5, 595–604.

Whatmore S. 2006. Materialist returns: practicing cultural geography in and for a more-than-human world. *Cultural Geographies* 13, 600–69.

Wisner B. 1993. Disaster vulnerability: scale, power, and daily life. *GeoJournal* 30:2, 127–140.

Wisner B., Blaikie P., Cannon T. and Davis I. 2004. *At Risk: Natural Hazards, People's Vulnerability, and Disasters* 2nd Edition. London: Routledge

Wolch J., Pincetl S. and Pulido L. 2001. Urban nature and the nature of urbanism. In: M. Dear (ed.) *From Chicago to L.A.: Making Sense of Urban Theory*. London: Sage, 367–402.

Zimmerer K.S. and Basset T.J. 2003. Approaching political ecology: society, nature and scale in human-environment studies. In: K.S. Zimmerer and T.J. Basset (eds) *Political Ecology: An Innovative Approach to Geography and Environment–development Studies*. New York: Guilford Press, 1–25.

Chapter 3

The Neoliberal Production of Vulnerability and Unequal Risk

Timothy W. Collins and Anthony M. Jimenez

Introduction

In 2006, the transnational metropolis of El Paso-Ciudad Juárez experienced several extreme precipitation events that led to widespread flood damage. Flood impacts were highly uneven within and between the cross-border communities. Impacts in Juárez were more severe than in El Paso, with about 5,000 homes damaged or destroyed, 20,000 residents left homeless, and extensive loss of built environmental infrastructure. Monetary damage in Juárez exceeded US$600 million, more than twice the city's annual budget. The most debilitating flood losses in Juárez were experienced by marginal residents of slums located within the city's rugged western *arroyos* (i.e., intermittently flowing landforms) (Collins 2009, 2010). In El Paso, an estimated 1500 homes and some flood protection infrastructure and roadways were damaged or destroyed by the floods, with recovery cost estimates of US$200 million. If monetary damage estimates on opposite sides of the border are compared, accounting for the sharp disparities in economic productivity, it is clear that the severity of the disaster in El Paso was at least an order of magnitude less than that experienced in Juárez. Still, flood impacts were severe enough to warrant a federal disaster declaration in El Paso, where recovery has been most difficult for socially marginal residents of central city *barrios* and peri-urban *colonias* (i.e., informal unincorporated settlements) (Collins 2009, 2010).

This chapter aims to expand political–economic understanding of the production of vulnerability and unequal risk – highlighted by events such as the 2006 El Paso-Ciudad Juárez floods – through consideration of recent advancements detailing the role of neoliberalization in generating environmental injustices. We begin by engaging political–economic studies of vulnerability to hazards. We then outline how the circulation of risk is integral to the circulation of capital and how, under neoliberalization, the power to direct the social surplus managed by states has been harnessed to facilitate accumulation for elites, with less powerful groups being marginalized and environmental injustices being amplified in the process. Next, we sketch three cases in which neoliberalization is implicated in the production of vulnerability and unequal risk: the emergence of disaster capitalism, the transfer of technological risks from global North-to-South, and the peri-urbanization of vulnerability in the global South. We then discuss the merits of a marginalization/

facilitation frame for explaining the neoliberal production of vulnerability and unequal risk. We argue that, to expand this frame, scholars must specify how neoliberal capital accumulation and risk transference strategies are technically conducted to engineer/resolve crises, avert confrontation, and facilitate movement across space, time, institutions, and social groups. We conclude by considering how these insights could inform vulnerability reduction.

The Political Economy of Vulnerability and Unequal Risk

A political–economic perspective on hazards provides a foundation for describing uneven patterns of risk, like those exposed by the 2006 El Paso-Ciudad Juárez floods (Wisner et al. 1976; O'Keefe et al. 1976; Susman et al. 1983; Watts 1983; Hewitt 1983; Bolin and Stanford 1998; Wisner et al. 2004). From this perspective, *risk* is viewed as the product of people's *exposure to an environmental hazard* and their *social vulnerability* (i.e., their capacity to anticipate, respond to, and recover from exposure to a chronic stressor or perturbation) (Wisner et al. 2004).

A premise of this chapter is that, while scholarship on hazard vulnerability offers insights for understanding patterns of unequal risk, it provides an incomplete basis for examining contemporary generative processes (Collins 2009, 2010). This is because it is not fully informed by developments in political–economic theory and analysis that reflect integrated changes in economic policies, technological systems, institutional arrangements, and demographic processes, all of which shape contemporary experiences of vulnerability. Embedded in dominant discussions of vulnerability and resilience (focused on global environmental change) are assumptions that prevailing political–economic processes are intended to generate aggregate economic growth, develop the global South, and improve human well-being (see Cannon and Müller-Mahn 2010). In such discussions, vulnerability is typically framed as a paradoxical, unfortunate by-product of globalization. Empirical analysis and theorizing about the neoliberal project urge us to reject such assumptions. Rather than providing the basis for understanding and challenging environmental injustices, such assumptions inadvertently perpetuate hegemonic discourses about the neutrality of neoliberalism and undermine the potential for analyses and interventions to escape the neoliberal frame.

Harvey (2005:2) refers to neoliberalism as "a theory of political economic practices that proposes that human well-being can best be advanced by liberating individual entrepreneurial freedoms and skills within and institutional framework characterized by strong private property rights, free markets, and free trade." The actual record of neoliberalization in terms of stimulating broad-based economic growth and advancing human well-being, according to Harvey (2005:156), is "nothing short of dismal." The redistribution of wealth from working-classes to elites and increasing social inequality have been such persistent features of neoliberalization the world over as to be regarded as fundamental aims of the entire project. In other words, neoliberalization in practice has been guided by the goal of restoring elite class power at the expense of less fortunate groups. And, if the neoliberal project is evaluated with that goal in mind, Harvey (2005:156) concludes that it "has been a huge success." The neoliberal turn has facilitated the concentration of wealth and power worldwide, successfully restoring class power to elites (as in the US and Britain), while in other cases creating conditions for capitalist class formation (as in China, India, and Russia). At the other end of the social spectrum, there has been a nearly "universal tendency to … expose the least fortunate elements in any society – be it Indonesia, Mexico, or Brazil – to the chill winds of austerity and the dull fate of increasing marginalization" (Harvey 2005:118). Here we consider what these political–economic transformations mean for understanding the production of vulnerability and unequal risk.

The Circulation of Capital and Risk

The Marxian concept of a social surplus is crucial to conceptualizing how vulnerability and risk are produced (or reduced) under the social relations of capitalism. Surplus value is generated through the exploitation of labor, and it can be redistributed to reduce vulnerabilities and risks through investment in three realms: fixed capital, the consumption fund, and social infrastructures. *Fixed capital* includes features of the built environment – created through investment of the social surplus – that are used to produce surplus value (e.g., streets). The *consumption fund* refers to commodities within the built environment that serve as means of consumption and which are necessary for social reproduction (e.g., parks). The term built environment, as used here, is a socially produced landscape composed of use values for production, exchange, and consumption (Harvey 2006:233). *Social infrastructures* regulate the capital–labor relation, provide the means to produce scientific and technical knowledge, contribute to the reproduction of labor power (through health care, education, social services, etc.), and offer means of ideological control and forums for dissent (Harvey 2006:398–399).

Investments in fixed capital and the consumption fund (which together form the built environment) can reduce physical exposure to hazards while investments in social infrastructures can reduce social vulnerability (Collins 2010). Fixed capital and consumption fund elements may be engineered, and the built environment modified, to protect human settlement and reduce hazard exposure. Specific social infrastructures to reduce risks include: technical assessment of hazards, land-use controls to restrict development in hazardous zones, warning systems and emergency planning, insurance to compensate losses, and post-event relief to enable recovery. More broadly, social infrastructures to reduce vulnerability include labor laws and access to safe housing, health care, education, and social services because such entitlements increase people's livelihood security, their material safety, and their coping capacities.

While fixed capital, the consumption fund, and social infrastructures are essential to the reproduction of capital and labor as well as the reduction of risk, there are barriers to individual capitalists' investing in their long-term formation and circulation. Such large-scale investments impose tremendous burdens since money has to be amassed to cover the initial costs, while the benefits are uncertain and potentially diffuse (Harvey 2006:264). Thus, expenditures have historically depended upon investment by the state. In other words, the state uses taxes as backing to undertake investments in the built environment and social infrastructures "which individual capitalists are unable or unwilling to furnish, however vital they may be as conditions of further accumulation" (Harvey 2006:395, 448). In turn, the state becomes a crucial field for struggle over expenditures in the built environment and social infrastructures. Thus, capitalists, working classes, and other interests must collectively constitute themselves and engage the state in a struggle for the power to shape the flow of public expenditures to achieve their desired aims. The unpredictable outcomes of such class struggles, in turn, have a direct bearing upon the production/reduction of vulnerability and risk.

The Circulation of Capital and Risk Under Neoliberalism

Harvey (2005) provides a basis for understanding the neoliberal circulation of capital and risk. Prior to neoliberalism, the period of "embedded liberalism" associated with Keynesian economic policies delivered high rates of economic growth in the advanced capitalist countries from World War II through the 1960s. Redistributive policies, controls on the mobility of capital, expanded public expenditures and welfare provisions, state interventions in the economy, and some planning of development accompanied relatively high rates of growth. During this period, "the state ... became a force field that internalized class relations. Working-class institutions such as labor unions and political parties of the left had a very real influence within the state apparatus" (Harvey 2005:11–12).

The crisis of capital accumulation in the 1970s, however, affected everyone through rising unemployment and inflation. A sense of discontent was widespread. The increased activity of socialist movements seemed to point towards the viability of a radical alternative, which threatened economic elites everywhere. Capitalists "had to move decisively if they were to protect themselves from political and economic annihilation" (Harvey 2005:15). Violent takeovers of state apparatuses (as in Chile and Argentina), abetted by domestic elites with US imperial support, relied on the use of repression and torture, demonstrating that brute force could be used to restore elite class power. Neoliberalization in countries such as the US and Britain, however, could only be accomplished by consensual means. Since a political movement explicitly aimed at the restoration of economic power for elites at the expense of all others would not gain much popular traction, the project needed to be ideologically legitimated, which the utopian discourse of neoliberal theory served well. The expressed aim of promoting individual freedoms, liberties, and rights held broad popular appeal and disguised the basic goal to restore class power. Since the 1980s, economic elites have re-oriented state powers to concentrate wealth in their own hands and they have simultaneously assumed a large degree of control in shaping sociospatial patterns of vulnerability and unequal risk.

What are some of the ways in which capital and risk have circulated unevenly under neoliberalism? Generally, wealth (i.e., tribute) has been siphoned from the global South-to-North, while financial risks have been pumped in the opposite direction. The financial risks imposed on global South states, in turn, have been leveraged, through the advent and implementation of structural adjustment programs, to pillage domestic economies while generating widespread social vulnerability and hazard exposure. The coercive structural adjustment plans mandated by international financial institutions (IFIs) in the aftermath of economic crises have served to fundamentally restructure societies in ways that amplified vulnerabilities and risks, especially for marginal groups in the global South (Davis 2006; Hamza and Zetter 1998; Keys et al. 2006; Wisner 2004).

What is important about this, in terms of theorizing about the circulation of risk, is the contradictory manner in which risk has been organized and assigned, and how risks have been translated from the abstract financial realm into intense forms of human suffering. In this system, there is a critical difference between neoliberal *theory* and *practice* in terms of the allocation of risk. In theory, lenders internalize risks that arise from bad investment decisions; yet, in practice, lenders have been facilitated in externalizing investment risks, while borrowers have been forced to repay debt to the detriment of ordinary people. The core neoliberal states gave the International Monetary Fund (IMF) and the World Bank (WB) full authority in 1982 to negotiate debt relief, in effect protecting investment banks against the risk of default. This enabled finance capitalists to extract high rates of return from the rest of the world during the 1980s and 1990s. International financial institutions (IFIs) have facilitated the transference of risks from the appropriate bearers – the lenders (i.e., the investment banks) – to the borrowers (i.e., the global South states for which the loans were purportedly intended to "develop"), enabling capital to be siphoned in the opposite direction. Debt crises in global South states – rare during the 1960s – became common. This system could not have been better designed to generate financial crises in debtor countries, paving the way to ransack capital and produce widespread vulnerability.

In sum, the restoration of power to economic elites (primarily in the global North) has drawn heavily on surpluses extracted from the rest of the world through international flows and structural adjustment practices. This contradicts neoliberal theory and is predicated on the concomitant transference of financial risks, which have eventually manifest as material risks (via increasing social vulnerability and hazard exposure), to the populations of global South states.

Neoliberalization has increased vulnerability to hazards across a variety of scales (Wisner et al. 2004), while contributing to environmental degradation (Heynan et al. 2007). As Harvey (2005:168) states "Neoliberalism seeks to strip away the protective coverings that embedded liberalism allowed and occasionally nurtured" through deregulation, privatization, and the withdrawal of the state's social safety net. Increasing social vulnerability has been produced in part through a general attack against labor, which has rendered "the disposable worker" archetypal on the neoliberal world stage (Wright 2006). Women and children often bear the worst risks (Cutter 1995). Social protections (e.g., pensions; health care; protections against injury and exposure; emergency and disaster response), which were formerly an obligation of employers and the state, are replaced by a "personal responsibility system" (Harvey 2005). Individuals buy products in markets that sell social protections instead. Security is thus privatized, becoming a matter of individual choice tied to the affordability of products embedded in risky financial markets (Klein 2007). The resulting system, which offers safety to the highest bidder, effectively strips marginal social groups of their "protective coverings," exposing them to myriad risks (Davis 2006). The consequences of neoliberalization for the degradation of protective social infrastructures, the amplification of social vulnerability, and the production of unequal risk are extreme, and discussed by way of example in the next section.

Given that the key achievement of neoliberal globalization has been to redistribute wealth to elites (rather than to generate broad-based development), it should come as no surprise that there has been a general lack of investment in fixed capital and consumption fund elements of built environments in many locales. These disinvestments have translated into unevenly increasing levels of hazard exposure (and disaster impacts) for certain communities (Wisner et al. 2004). Public infrastructure systems the world over have been privatized, excluding the least fortunate from fulfilling basic needs (e.g., clean water), or have been left to decay, which has heightened vulnerability and amplified unequal risks (Collins 2010; Davis 2006; Graham 2010; McFarlane 2010; Swyngedouw 2004; Verchick 2010). It is important to note that, while neoliberalization has generated the most acute vulnerabilities and risks in the global South, the uneven degradation of fixed capital, the consumption fund, and social infrastructures has also produced disaster vulnerability in the North (ASCE 2005; Collins 2009; Comfort 2006; Freudenburg et al. 2008).

From a Marxist perspective, the adept usage of legitimating discourses and the technical complexity of financial markets that define neoliberalization conceal what is fundamentally an historical–geographical epoch marked by capitalists' use of *primitive accumulation* as a favored enrichment strategy (Harvey 2005). *Accumulation by dispossession* has relied on the expansion of accumulation practices which Marx treated as "original" to the rise of capitalism. By definition, accumulation by dispossession produces vulnerability and amplifies unequal risks. Thus it could alternatively be termed *accumulation by endangerment*, which has expanded on several fronts via neoliberalization. First, financial "innovations" made the credit system an important field for facilitating accumulation by dispossession through speculation, predation, fraud, and theft. Second, through privatization of hitherto public assets, new arenas for capital accumulation in domains previously regarded as beyond the calculus of profitability – including public utilities, social welfare provision, and public institutions – have been successfully opened throughout the world. Third, once neoliberalized, states have aggressively promoted capitalist elites' accumulation strategies, facilitating dispossession and endangering ever larger population segments worldwide. Next, we discuss these processes by way of three case examples.

Neoliberalization and the Production of Vulnerability and Unequal Risk: Three Trends

The neoliberal production of vulnerability and unequal risk is exemplified in three trends: (1) the emergence of disaster capitalism as an accumulation strategy, (2) the transfer of technological risks from global North-to-South, and (3) the peri-urbanization of vulnerability in the global South.

The Rise of Disaster Capitalism

Hazards scholars have analyzed the role of rewards associated with hazards in the production of vulnerability. Pelling (1999), e.g., incorporated a consideration of the economic rewards associated with environmental hazards in his analysis of unequal risk. He clarified the manner in which Guyanans viewed the international resources that domestic flood hazards attracted as "a potential source of rents for [state] institutions ... as well as for individuals and private sector entrepreneurs" (Pelling 1999:258). Due to pre-existing power differentials, the economic benefits experienced by marginal groups within the informal economic sector were meager, while elites co-opted the lion's share of rewards. The uneven allocation of flood hazard resources – intended to reduce vulnerability – ultimately reproduced "embedded distributions of power and vulnerability" (Pelling 1999:249). Other work demonstrates that people associate economic and environmental rewards with hazards, and that such benefits influence the production of vulnerability and unequal risk (Collins 2008, 2009, 2010). A shared assumption of these studies, however, is that economic rewards are *associated with* hazards and disasters.

Recent scholarship has focused on how environmental disasters (like other catastrophes, such as engineered debt crises and wars) have become *structurally central* to political–economic processes and accumulation strategies under neoliberalism (Klein 2007; Fox Gotham 2008; Gunewardena and Schuller 2008). Klein (2007:6) terms the "orchestrated raids on the public sphere in the wake of catastrophic events, combined with treatment of disasters as exciting market opportunities, 'disaster capitalism'." As the limits to neoliberalization for broad-based development become increasingly apparent and opposition to neoliberal policies becomes more entrenched (as illustrated by the Zapatista uprising, anti-IMF riots, and the anti-globalization movement), agents of neoliberalism have become adept at generating and exploiting crises as a means to implement their accumulation by dispossession and endangerment strategies.

For example, following the 2004 South Asian Tsunami the Sri Lankan state designated the coastline as a "high risk" environment in order to justify its expropriation from fishing communities for global ecotourism development, leading to the loss of land and livelihood among locals (Gunewardena 2008; Klein 2007). Prior to the tsunami, IFIs and domestic economic elites had presented plans for exclusive, privatized ecotourism development along Sri Lanka's coast. These plans were deeply unpopular and had been scrapped by Sri Lankan voters. The tsunami, however, provided an opportunity for elites to push the plan's unpopular neoliberal agenda through under "emergency" cover. After implementing "crisis" rules (e.g., no reconstruction of fishing villages near the beach), the Sri Lankan state subsidized a contradictory round of coastline "development," in which they authorized the construction of exclusive beachside resorts, thus facilitating a form of accumulation – associated with the consumption of coastal amenities by global elites – founded on the dispossession of local fishers. Klein (2007:401–402) concludes that, because the tsunami so effectively cleared the beach,

> ... a process of displacement that would normally unfold over years took place in a matter of days or weeks. What it looked like was hundreds of thousands of poor, brown-skinned people (the fishing people deemed "unproductive" by the WB) being moved against their wishes to make room for ultra-rich, mostly light-skinned people (the "high-yield" tourists).

This neoliberal outcome – involving the facilitation of accumulation by elites at the direct expense of the marginalization of ordinary residents – has been repeated in the elsewhere in the aftermath of the tsunami (Gunewardena 2008; Klein 2007) and other recent disasters, including Hurricanes Mitch (Stonich 2008) and Katrina (Adams et al. 2009; Fox Gotham 2008; Klein 2007; Reed 2006), and 9/11 (Damiani 2008; Fox Gotham 2008). Thus, based on recent disaster experiences, one can conclude that, while elite groups may choose to expose themselves and their investments to risks as they pursue opportunities associated with hazards, crises and catastrophes, neoliberal institutional arrangements have increasingly enabled them to transfer risks to less powerful people and to expropriate rewards.

Is this Technology Transfer or the Transference of Technological Risk?

In late modernity (post-World War II) the industrialization process fueling the growth of advanced capitalist societies has been predicated on the production of new "technological" risks (Beck 1992). These sorts of hazards were not treated in classical Marxist thought, but they must be regarded as integral to the uneven circulation of capital and risk in late modernity.

Recent scholarship has mapped the transference of toxic risks from the global North-to-South (Adeola 2000; Clapp 2001; Frey 2003; Pellow 2007), and emerging trend that can be understood via the "treadmill of production" framework (Gould et al. 2008). From this perspective, capitalist economies by design create ecological and social harm through a self-reinforcing mechanism of increasing rates of production and consumption. Basic contradictions undermine the conditions for production through environmental degradation and/or damage to the health of workers, to the point at which this spills over into the social arena, leading to mass resistance from environmental and/or labor movements (O'Connor 1988). This happened during the later years of embedded liberalism, when ecological degradation aroused popular consciousness in the global North, spawning environmental movements that made progress in pushing through state regulatory reforms (back when the state acted more like a force field that internalized class conflict) (Gottlieb 2005; Szasz 1994).

To maintain an increasing rate of return on investments, when environmental resources become limited and/or oppositional movements emerge, the capitalist "treadmill" searches for new spaces to exploit rather than restructuring production along "sustainable" lines. Indeed, the success achieved by northern movements – in tandem with neoliberal governance arrangements – led many global North-based corporate polluters to move their hazardous production and waste-disposal operations to the global South for less controlled and more profitable pastures (under the guises of development, recycling, charitable donation, and technology transfer). The political success of the environmental justice movements in the US, e.g., has contributed to the flow of toxic risk to the South and, in effect, has inadvertently helped rescale patterns of environmental injustice in exposure to toxics from the domestic to the transnational level (Goldman 1996; Low and Gleeson 1998; Pellow 2007).

In 1988 there were 6,200 medical waste incinerators in the US; by 2003 that number had dropped to 115. Due to vociferous NIMBY-based activism, incineration is dying as a waste disposal technology in the North. Meanwhile, incineration "technology transfer" to the global South is being facilitated by IFIs such as the WB (Pellow 2007).

Consider another example. Today many pesticides banned for use in the US and other northern states are exported, dumped or used throughout the global South (Smith et al. 2008). Regulation of pesticides in global North states (e.g., the 1973 US ban on DDT) is the fruit of work by environmentalists and farmworker advocates to bring attention to the intense human and ecological harm produced by these toxics. Such regulation has not solved the problem, but rather displaced it. Pellow (2007:156) describes how neoliberal arrangements promoted by IFIs have worsened global pesticide injustices:

> Such arrangements often bring a large volume of pesticide use to these nations because the export-oriented agricultural framework of most international loan programs is built around the production of crops ... which require more pesticides than food crops for domestic consumption. These agreements provide new and growing markets for agrochemical corporations, particularly since the human health and environmental impacts of pesticide pollution are becoming known in the North, reducing their popularity and introducing new restrictions

and bans on certain chemicals ... Between 1988 and 1995, the World Bank
financed the purchase of $350.8 million worth of pesticides ... which were then
sent to southern nations as part of international development aid packages.

The environmental injustices associated with pesticide use are widespread,
embedded in global sociospatial hierarchies, and observable across race, class,
gender, age, migration status, and nation. While a small group of transnational
corporations control the profits of the global pesticide market, the health risks of
pesticide exposure tend to burden the people who benefit least from pesticide use
(Frey 1995; Garcia 2003; London et al. 2002).

From a Marxian perspective, the global transference of technological risk
exemplifies the deployment of the *spatial fix* in late modernity, as it provides
a solution to the toxic crisis of capitalism experienced in the North through
geographic expansion to the South (Harvey 2006); this is what Pellow (2007)
terms the *hyperspatiality of risk*. In this case, to ensure the viability of capital
accumulation within an unsustainable and unjust political economy in the face
of growing resistance to socio-ecological harm (in the global North), toxic
hazards are being displaced to other world regions (where new ecosystems and
social groups are being polluted). Like others, we argue that these toxics should
be conceived of as forms of "anti-use-value" or "anti-wealth" (Low and Gleeson
1998; Frey 2003). Hence, sending toxics to the global South not only adheres to
the pattern of siphoning wealth out of former colonies, it represents a new form
of exploitation as it involves exporting substances that function as the opposite
of wealth, draining local resources and capacities to produce into the future.
Since the transference of toxic risks to the global South poisons conditions of
production (robbing people and places of their future prospects), it can be viewed
as a late modern strategy for accumulation by dispossession and endangerment. A
fundamental aspect of this neoliberal system of flows, involving the circulation of
capital and toxic risk, is its inherently unequal nature. It permits powerful people
and states to accumulate capital while displacing toxic risks to less powerful social
spaces, where vulnerable people must cope with harmful exposures.

The Peri-urbanization of Vulnerability in the Global South

While peasant producers in the global South have been forced to sink-or-swim
in global commodity markets – with their existences becoming increasingly
toxic and precarious as a result – the concentration of social vulnerability and
hazard exposure in the global South under neoliberalization has materialized
most substantially via the proliferation of peri-urban slums (UNHSP 2003).
Davis (2006) documents the role of neoliberalization in producing vulnerability
– primarily through the desperate rural-to-urban movement of human bodies –
in slums of the global South. Mass slum formation has been driven less by the
pull of global South cities, many of which have been economically weakened
via neoliberalization, and more by forces pushing people form the countryside.

Social safety nets have disappeared and individual farmers have become increasingly vulnerable to myriad perturbations (e.g., drought, inflation, rising interest rates, falling commodity prices), which eventually force them to migrate to cities. Meanwhile, the privatization of once commonly held lands – a central aim of neoliberalization – has led to analogous prospects for the peasantry. The privatization of the *ejidos* in Mexico during the 1990s, e.g., forced many rural dwellers off the land into the cities (Harvey 2005).

The mass rural-to-urban migration of human bodies in the global South represents a kind of grassroots spatial fix to the crisis of social reproduction wrought by neoliberalization, a "solution" that has directly translated into the problem of slum formation. Attempts to address the "overurbanization" problem in the global South have been unsuccessful, largely due to institutionalized power asymmetries that favor elites to the detriment of the poor and powerless. Public housing and WB slum upgrade programs – initially justified to support the housing needs of the poor – have all too often been poached by middle and elite classes (Berner 2000; Leonard 2000; O'Hare et al. 1998; Peattie 1987; Pezzoli 1995). Regressive tax structures – promoted by the IMF and WB – have relied on user fees and charges for public services rather than tapping into the income, conspicuous consumption, and real estate of the wealthy (Devas 2003). A discourse of "self-help" has also been promulgated by states, the IMF and WB, and non-governmental organizations (NGOs) to legitimate the abdication of any commitment to intervene in the structural causes of poverty and exclusion (Seabrook 1996). Meanwhile, no effort has been made to counteract the negative effects of rampant land speculation by the wealthy (domestically in the global South, urban real estate is viewed as the most secure capital shelter to protect against the currency devaluations produced by neoliberalization), which include the monopolization and concentration of land ownership, decreasing land availability, and soaring rents that exclude the poor from the urban housing market (Bankoff 2003; López-Morales 2010). In this context, the neoliberal hype regarding the benefits of slum privatization via land titling appears short-sighted, since "[t]he very market forces ... that the WB currently hails as the solution to the Third World urban housing crisis are the classical instigators of that same crisis" (Davis 2006:93). In sum, assistance efforts in the global South have tended to benefit the well-heeled minority at the direct expense of the increasingly impoverished masses. As much as 80 percent of the housing demand in global South cities is supplied informally (UNHSP 2003).

A picture emerges of sacrificial social spaces defined by their extreme marginality and vulnerability, where hazards abound and access to basic public health resources – including the consumption fund and social infrastructures – is denied (Bankoff 2003; Hamza and Zetter 1998; McFarlane 2010; Mustafa 2005). The lack of access to basic sanitation infrastructure coupled with exposure to urban pollution means that a double burden of disease is often shouldered by slum dwellers. Sewage is typically untreated, which pollutes water supplies, condemning droves of people – children especially – to die from preventable diseases. Malnutrition abounds. Yet these people also suffer from the spectrum of chronic diseases associated with industrialization. Meanwhile, the poor have been systematically excluded from healthcare. "The neoliberal restructuring of Third World urban economies ... has had a devastating impact on the public provision of healthcare ... the coerced tribute that the Third World pays to the First World has been the literal difference between life and death for millions of people" (Davis 2006:147–148).

Thus, informal peri-urban settlements embody many of the contradictions of the neoliberal political economy. These social spaces are produced by global power asymmetries, reflect acute vulnerability, and are unjustly burdened by risks. Despite these injustices, slums will continue proliferate in the global South into the foreseeable future, since they offer the only housing options accessible to the multitudes displaced via neoliberalization.

Marginalization, Facilitation, and Neoliberal Risk Translation

We emphasize a few points with regard to the role of global political–economic processes in the production of vulnerability and unequal risk. First, in examining the political–economic context of vulnerability, scholars should acknowledge that neoliberalism, despite the progressive rhetoric surrounding it, has, in practice, been a project for the restoration of elite class power predicated on dispossession, marginalization, and endangerment. Over the past 30 years, capitalist elites have channeled state powers to concentrate wealth in their own hands and have simultaneously *produced* patterns of vulnerability and unequal risk. As Harvey (2006:xxvi) observes: "The genius of the current structure of institutions is not only to spread risks, but also to spread them asymmetrically in such a way as to ensure that the costs ... are visited for the most part on those least able to afford them."

These asymmetrical–relational dynamics have been accounted for theoretically and empirically in analyses of the production of vulnerability and unequal risk. Collins (2008, 2009, 2010) introduced a "marginalization/facilitation frame" to expand upon critical hazards concepts and explain these processes in cases such as the ones described above. Within the hazards literature, the concept of marginalization is the best known explanation for the production of vulnerability. *Marginalization* connotes how less powerful sociospatial groups are made vulnerable as institutional arrangements limit their livelihood options (i.e., dispossess them), pressuring them to degrade landscapes and occupy hazardous environments and labor positions, while they experience decreased capacities to cope with socio-ecological change (Blaikie and Brookfield 1987; Susman et al. 1983). In political–economic terms, understanding vulnerability production necessitates conjoint analysis of the capital accumulation strategies that underpin marginalization. *Facilitation* is a relational term referring to how the same (neoliberal) institutional arrangements enable powerful geographical groups of people to externalize risks (by shifting costs to other people, places, and/or times) and capitalize on opportunities associated with hazardous environments and production processes (Collins 2008, 2009, 2010). As our discussion above emphasizes, vulnerability formation under neoliberalism has been driven by the expansion of strategies for accumulation by dispossession and endangerment. Because experiences of marginalization are predicated on facilitation, the marginalization/facilitation framework expands analytical focus beyond pressures that appear to directly amplify vulnerability to clarify the systemic political–economic dynamics that facilitate accumulation, a leap that must be made to better understand the production of vulnerability and unequal risk.

Second, we argue that scholars must recognize and attempt to clarify how capital accumulation and risk transference strategies are *technically conducted* – in terms of how they are orchestrated and channeled – to engineer/resolve crises, avert confrontation, and facilitate movement across space, time, institutions, and social groups. As noted above, the spatial fix provides one strategy for overcoming contradictions and resolving accumulation crises, e.g., by shifting toxic industrial operations into poor minority neighborhoods or to the global South. These neoliberal technologies, which are used to facilitate capital accumulation and to transfer risk, however, have depended on much more than the displacement of things and their associated effects.

In terms of the technologies of risk transference under neoliberalism, we find it useful to conceive of displacement in more-than-material terms. In the neoliberal order, states and IFIs have served first and foremost as *vehicles* or *media for conveying power* to economic elites, insofar as these global social infrastructures have been designed to legitimate and facilitate capital accumulation through the transmutation of risk from material form (e.g., the exploitation of labor and nature to extract surplus value) to symbolic equivalent (e.g., dollars, interest rates, speculative gains) and back again (e.g., debt-servicing and structural adjustment mandates to dispossess labor). Thus, the transmission of capital and risk depends on a series of material–symbolic metamorphoses that can only be accomplished through the coordination of neoliberalized social infrastructure networks spanning the globe. Neoliberal risk translation technologies are simultaneously impersonal and extremely invasive at a bodily level, enabling capital to relentlessly dismantle all barriers to accumulation – with extreme social and ecological consequences – nearly without a trace. The shape-shifting fluidity that characterizes the circulation of capital and risk under neoliberalism constrains attempts to explain vulnerability production processes in reference to rigid conceptions, and confounds efforts to account for metabolic flows of capital and risk in a manner amenable to precision analysis. Nonetheless, developing more systematic knowledge of neoliberal risk translation technologies is critically important for explaining of the production of vulnerability. We assert that neoliberal risk translation technologies may be brought into clearer focus by recognizing the built environment and social infrastructures as "thoroughly political constructions which tend to embody 'congealed social interests'" rather than as the mere technical domains of experts such as engineers and economists (Graham 2010:13).

Conclusions

We conclude by considering a normative question. In this context, *how might we limit the production of vulnerability while generating more just socioenvironmental arrangements?* First, taking into account neoliberal technologies of risk displacement, environmental justice advocates must recognize that, in their efforts to reduce local risks, they may inadvertently help transfer risks elsewhere – to less powerful social spaces. Thus we echo the call for the abandonment of a parochial socioenvironmental justice ethic (i.e., not in my backyard or NIMBY) and the adoption of a cosmopolitan one (i.e., not on planet earth or NOPE) as a necessary step toward locating more successful vulnerability reduction approaches (Low and Gleeson 1998; Pellow 2007).

Second, an environmental justice frame to successfully promote vulnerability reduction must seek to isolate and redress a largely hidden issue: the neoliberal state's reliance on "cost-benefit analysis" for decision-making about the allocation of the social surplus. Cost-benefit analysis is ubiquitously used by the social infrastructures of the neoliberal state as a technical "science-based approach" for allocating resources (and risks). In the US, cost-benefit analysis has clear neoliberal provenance. It was institutionalized early in the Reagan presidency as an instrument of deregulation (Harvey 2005:52). Theoretically, cost-benefit analysis is rooted in utilitarianism: "The moral theory that judges the goodness of outcomes – and therefore the rightness of actions as they affect outcomes – by the degree to which they secure the greatest benefit of all concerned" (Hardin 1988:21). Thus cost-benefit analysis might provide a rational basis for allocating state funds to reducing vulnerability among marginal groups, if such an outcome were defined in terms of "the greatest benefit for all concerned."

However, under neoliberalism, there are at least two reasons why cost-benefit analysis has enhanced injustice by further entrenching sociospatial inequalities, increasing vulnerability among marginal groups, and amplifying unequal risks. One, within the neoliberal state – to the degree to which utilitarianism is operationalized based on business activity, property valuation (for tax assessment) and the like – there is a strong tendency for "successful" allocation "actions" to facilitate the already wealthy while further marginalizing the poor. Cost-benefit analysis, e.g., is routinely employed to justify allocating social protection resources for elites living in hazardous environments with high exchange value and to simultaneously warrant the dislocation of socially-marginal occupants of landscapes with low exchange value. Two (and related), in the context of dramatic "background injustice" (Rawls 1993) – itself amplified by neoliberalization – cost-benefit analysis only serves to increase asymmetric power relations and socioenvironmental injustices through time. Nonetheless, cost-benefit analysis provides the technocratic managerial basis for allocating risk reduction funding by the neoliberal state. For example, the US Federal Emergency Management Agency's (FEMA) Hazard Mitigation Grant Program and Flood Mitigation Assistance Program mandate that any mitigation activities funded meet "cost-effectiveness" criteria.

The local state's implementation of disaster recovery programs following the 2006 flood disaster in El Paso-Ciudad Juárez exemplifies the injustices embedded within the cost-benefit calculus (Collins 2009, 2010). Within the City of El Paso, investment of the social surplus by the local state in flood recovery was uneven, facilitating elite geographical groups of people on the city's "Westside" at the expense of less powerful sociospatial groups. In accessing resources for flood recovery, the Westside was privileged to have highly educated residents who drew on technical knowledge to place specific demands for remediation on the local state, which were met by City officials. In cases of Westside civil engineering failure and flood damage, the City of El Paso deemed fixed capital upgrades to be "feasible." Many Westside flood locations were placed at the top of the City's priority list for infrastructure improvement. In contrast, the City marginalized less powerful geographical groups through the application of cost-benefit analysis in its "hazard reduction" programs. For example, two socially marginal neighborhoods experienced severe flooding in 2006. Rather than upgrade flood protection infrastructure, as was done on the Westside, the City chose to buy-out properties in the two neighborhoods, demolish homes, and relocate residents.

Why were built environmental improvements deemed feasible on the Westside and infeasible in less powerful neighborhoods? The cost-benefit approach used by the City advantaged powerful residents of high exchange value landscapes and disadvantaged occupants of property with relatively low exchange value. The higher exchange value of property on the Westside made investments to improve the built environment and reduce flood hazards economically rational. In contrast, in the two socially marginal neighborhoods with great recovery needs, the low exchange value of property served to rationalize public investment in residential displacement rather than in built environmental improvements. In other words, the City determined that it was more "cost-effective" to purchase/destroy home sites and relocate people than to pay to improve flood protection infrastructure. As a city council representative put it: "The buyouts in the Saipan area cost about $5 million, but fixing the storm-water management problem there would have cost about $10 million" (Collins 2009).

Cost-benefit analysis is ultimately guided by " … an unwritten code that justifies development for some at the cost of others" (Pellow 2007:141). This is a code reflected in neoliberal economic orthodoxy and practice. Take, e.g., this excerpt from the infamous internal WB memo circulated by Lawrence Summers (1991), then the chief economist and vice president of the WB, and now White House economic advisor:

> Shouldn't the WB be encouraging MORE migration of the dirty industries to the LDCs [lesser developed countries]? … A given amount of health impairing pollution should be done in the country with the lowest cost, which will be the country with the lowest wages. I think the economic logic behind dumping a load of toxic waste in the lowest wage country is impeccable and we should face up to that.

Thus, upon inspection, cost-benefit analysis is discernable as a state derivation of neoliberal economic theory and practice (as reflected in the above quote), and it produces powerful, seemingly objective, truth effects for legitimating relational processes of marginalization/facilitation. The espoused neutrality of cost-benefit analysis, moreover, has left it largely unchallenged as a neoliberal discourse for legitimating the unjust allocation of state resources. This must change if the social surplus managed by the state is to play a progressive role in reducing vulnerability. Without the implementation of genuine need-based programs to ameliorate vulnerability through the empowerment of marginal sociospatial groups, institutional resources are all too easily co-opted by the powerful – those least in need of assistance – which deepens environmental injustices while reproducing vulnerabilities.

References

Adams, V., Van Hattum, T. and D. English. 2009. Chronic disaster syndrome: displacement, disaster capitalism, and the eviction of the poor from New Orleans. *American Ethnologist* 36, 615–636.

Adeola, F. 2000. Cross-national environmental injustice and human rights issues – a review of evidence in the developing world. *American Behavioral Scientist* 43, 686–706.

Alexander, D. 1993. *Natural Disasters*. New York: Chapman and Hall.

ASCE (American Society of Civil Engineers). 2009. Report card for America's infrastructure. Retrieved from: <http://www.asce.org/reportcard/>. Accessed November 22, 2010.

Bankoff, G. 2003. Constructing vulnerability: the historical, natural and social generation of flooding in Metropolitan Manila. *Disasters* 27, 224–238.

Beck, U. 1992. *Risk Society: Towards a New Modernity* (trans. Mark Ritter). London: Sage.

Berner, E. 2000. Poverty alleviation and the eviction of the poorest: towards urban land reform in the Philippines. *International Journal of Urban and Regional Research* 24, 536–553.

Blaikie, P. and H. Brookfield. 1987. *Land Degradation and Society*. London: Methuen.

Blaikie, P., Cannon, T., Davis, I. and B. Wisner. 1994. *At Risk: Natural Hazards, People's Vulnerability, and Disasters*, first edition. London: Routledge.

Bolin, B. and L. Stanford. 1998. *The Northridge Earthquake: Vulnerability and Disaster*. London: Routledge.

Cannon, T. and D. Müller-Mahn. 2010. Vulnerability, resilience and development discourses in the context of climate change. *Natural Hazards* 55:3, 621–635.

Clapp, J. 2001. *Toxic Exports: The Transfer of Hazardous Wastes from Rich to Poor Countries*. Ithaca: Cornell University Press.

Collins, T. 2008. The political ecology of hazard vulnerability: marginalization, facilitation and the production of differential risk to urban wildfires in Arizona's White Mountains. *Journal of Political Ecology* 15, 21–43.

Collins, T. 2009. The production of unequal risk in hazardscapes: an explanatory frame applied to disaster at the U.S.–Mexico border. *Geoforum* 40, 589–601.

Collins, T. 2010. Marginalization, facilitation, and the production of unequal risk: the 2006 Paso del Norte floods. *Antipode* 42, 258–288.

Comfort, L. 2006. Cities at risk: Hurricane Katrina and the drowning of New Orleans. *Urban Affairs Review* 41, 501–516.

Cutter, S. 1995. The forgotten casualties: women, children, and environmental change. *Global Environmental Change* 5:3, 181–194.

Damiani, B. 2008. Capitalization of post-9/11 recovery. In: *Capitalizing on Catastrophe: Neoliberal Strategies in Disaster Reconstruction*, (eds) N. Gunewardana and M. Schuller, 157–172. Lanham: Alta Mira Press.

Davis, M. 2006. *Planet of Slums*. New York: Verso.

Devas, N. 2003. Can city governments in the south deliver for the poor? A municipal finance perspective. *International Development Planning Review* 25, 1–25.

Fox Gotham, K. 2008. From 9/11 to 8/29: post-disaster recovery and rebuilding in New York and New Orleans. *Social Forces* 87, 1039–1062.

Freudenburg, W., Gramling, R., Laska, S. and K. Erikson. 2008. Organizing hazards, engineering disasters? Improving the recognition of political–economic factors in the creation of disasters. *Social Forces* 87, 1015–1038.

Frey, R. 1995. The international traffic in pesticides. *Technological Forecasting and Social Change* 50, 151–169.

Frey, R. 2003. The transfer of core-based hazardous production processes to the export processing zones of the periphery: the Maquiladora centers of northern Mexico. *Journal of World Systems Research* 9, 317–354.

Garcia, A. 2003. Pesticide exposure and women's health. *American Journal of Industrial Medicine* 44, 584–594.

Goldman, B. 1996. What is the future of environmental justice? *Antipode* 28:2 122–141.

Gottlieb, R. 2005. *Forcing the Spring: The Transformation of the American Environmental Movement*, 2nd edition. Washington, DC: Island.

Gould, K., Pellow, D. and A. Schnaiberg. 2008. *The Treadmill of Production: Injustice and Unsustainability in the Global Economy*. Boulder: Paradigm.

Graham, S. 2010. When infrastuctures fail. In: *Disrupted Cities: When Infrastructure Fails*, (ed.) S. Graham, 1–26. London: Routledge.

Gunewardena, N. 2008. Peddling paradise, rebuilding Serendib: the 100-meter refugees versus the tourism industry in post-tsunami Sri Lanka. In: *Capitalizing on Catastrophe: Neoliberal Strategies in Disaster Reconstruction*, (eds) N. Gunewardana and M. Schuller, 69–92. Lanham: Alta Mira Press.

Gunewardena, N. and M. Schuller, (eds) 2008. *Capitalizing on Catastrophe: Neoliberal Strategies in Disaster Reconstruction*. Lanham: Alta Mira Press.

Hamza, M. and R. Zetter. 1998. Structural adjustment, urban systems, and disaster vulnerability in developing countries. *Cities* 15 291–299.

Hardin, R. 1988. *Morality Within the Limits of Reason*. Chicago: University of Chicago Press.

Harvey, D. 2005. *A Brief History of Neoliberalism*. New York: Oxford University Press.

Harvey, D. 2006. *The Limits to Capital*. London: Verso.

Hewitt, K. (ed.) 1983. *Interpretations of Calamity from the Viewpoint of Human Ecology*. Boston: Allen and Unwin.

Heynan, N., McCarthy, J., Prudham, S. and P. Robbins, (eds) 2007. *Neoliberal Environments: False Promises and Unnatural Consequences*. London: Routledge.

Keys, A., Masterman-Smith, H. and D. Cottle. 2006. The political economy of a natural disaster: the Boxing Day tsunami, 2004. *Antipode* 38, 195–204.

Klein, N. 2007. *The Shock Doctrine: The Rise of Disaster Capitalism*. New York: Metropolitan.

Leonard, J. 2000. City profile: Lima. *Cities* 17, 433–445.

London, L., de Grosbois, S., Wesseling, C., Kisting, S., Rother, A. and D. Mergler. 2002. Pesticide usage and health consequences for women in developing countries: out of sight, out of mind? *International Journal of Occupational and Environmental Health* 8, 45–59.

López-Morales, E. 2010. Real estate market, state-entrepreneurialism and urban policy in the "gentrification by ground rent dispossession" of Santiago de Chile. *Journal of Latin American Geography* 9, 145–173.

Low, N. and B. Gleeson. 1998. *Justice, Society and Nature: An Exploration of Political Ecology*. London: Routledge.

McFarlane, C. 2010. Infrastructure, interruption, and inequality: urban life in the global south. In: *Disrupted Cities: When Infrastructure Fails*, (ed.) S. Graham, 131–144. London: Routledge.

Mustafa, D. 2005. The production of an urban hazardscape in Pakistan: modernity, vulnerability, and the range of choice. *Annals of the Association of American Geographers* 95, 566–586.

O'Connor. J. 1988. Capitalism, nature, socialism: a theoretical introduction. *Capital, Nature, Socialism* 1, 11–38.

O'Hare, G., Abbott, D. and Burke, M. 1998. A review of slum housing policies in Mumbai. *Cities* 15 269–283.

O'Keefe, P., Westgate, K. and Wisner, B. 1976. Taking the "naturalness" out of "natural disaster." *Nature* 260, 566–567.

Peattie, L. 1987. Affordability. *Habitat International* 11:4, 69–76.

Pelling, M. 1999. The political ecology of flood hazard in urban Guyana. *Geoforum* 30, 249–261.

Pellow, D. 2007. *Resisting Global Toxics: Transnational Movements for Environmental Justice*. Cambridge: MIT Press.

Pezzoli, K. 1995. Mexico's urban housing environments: economic and ecological challenges of the 1990s. In: *Housing the Urban Poor: Policy and Practice in Developing Countries*, (eds) B. Aldrich and R. Sandhu, 140–165. London: Zed Books.

Rawls, J. 1993. *Political Liberalism*. New York: Columbia University Press.

Reed, A. 2006. Undone by neoliberalism: New Orleans was decimated by an ideological program, not a storm. *The Nation* 283:8, 26–30.

Seabrook, J. 1996. *In the Cities of the South: Scenes from a Developing World*. London: Verso.

Smith, C., Kerr, K. and Sadripour, A. 2008. Pesticide exports from U.S. ports, 2001–2003. *International Journal of Occupational and Environmental Health* 14, 167–177.

Stonich, S. 2008. International tourism and disaster capitalism: the case of hurricane Mitch in Honduras. In: *Capitalizing on Catastrophe: Neoliberal Strategies in Disaster Reconstruction*, (eds) N. Gunewardana and M. Schuller, 47–68. Lanham: Alta Mira Press.

Summers, L. 1991. World Bank internal memo, circulated 12 December 1991. Retrieved from: <http://www.whirledbank.org/ourwords/summers.html>. Accessed November 22, 2010.

Susman, P., O'Keefe, P. and Wisner, B. 1983. Global disasters, a radical interpretation. In: *Interpretations of Calamity from the Viewpoint of Human Ecology*, (ed.) K. Hewitt, 263–283. Boston: Allen and Unwin.

Swyngedouw, E. 2004. *Social Power and the Urbanization of Water: Flows of Power*. Oxford: Oxford University Press.

Szasz, A. 1994. *Ecopopulism: Toxic Waste and the Movement for Environmental Justice*. Minneapolis: University of Minnesota Press.

UNHSP (United Nations Human Settlements Programme). 2003. *The Challenge of Slums: Global Report on Human Settlements*. London: Earthscan.

Verchick, R. 2010. *Facing Catastrophe: Environmental Action for a Post-Katrina World*. Cambridge: Harvard University Press.

Watts, M. 1983. On the poverty of theory: natural hazards research in context. In: *Interpretations of Calamity from the Viewpoint of Human Ecology*, (ed.) K. Hewitt, 231–262. Boston: Allen and Unwin.

Wisner, B., Westgate, K. and O'Keefe, P. 1976. Poverty and disaster. *New Society* 9 547–548.

Wisner, B., Blaikie, P., Cannon, T. and Davis, I. 2004. *At Risk: Natural Hazards, People's Vulnerability and Disasters*, second edition. London: Routledge.

Wright, M. 2006. *Disposable Women and Other Myths of Global Capitalism*. London: Routledge.

Chapter 4

The Production of Urban Vulnerability Through Market-based Parks Governance

Harold A. Perkins

Introduction: (Re) Negotiating Provision for Urban Parks

Urban parks have long been designed to shape social relations under capitalism. Landscape parks were built in cities in the United States during the 19th century to acculturate restive working classes to middle and upper class values (Rosenzweig, 1983). Those sprawling green landscapes modeled after Victorian gardens were provided as leisurely alternatives to socializing in taverns (Rosenzweig and Blackmar, 1992). Many landscape parks were later remodeled during the Great Depression and actively programmed with recreational activities meant to be diversions for unemployed workers waiting for jobs and better economic times (Cranz, 1989). Parks in the prosperous postwar era were built with a wide array of active and passive elements to provide workers and their families places to retreat and rejuvenate from hectic city life (Jones and Wills, 2005). Parks directors today make difficult choices in programming as the diversity of parks patrons increased with the rise of immigration to the United States from regions like Latin America. As people move to the United States from around the world, they bring with them cultural expectations for green space that do not always mesh well with other established ethnic groups' preferences for parks programming. Parks programmers in many cities now try to figure out how they can manage park spaces as sites of multicultural interaction (Low et al., 2005).

These social engineering projects represent "negotiations" between municipal government, capital, and civil society concerning the (re)distribution of resources that enhance the standard of living in cities through contact with green space (Rosenzweig, 1983; Cranz, 1989; Perkins, 2009a). This is quite evident in the urban environmental history of Milwaukee, Wisconsin – a city well-known for its large urban park system. Laborers rioted there in early May of 1886 over working and living conditions in the city. The riots led industrialists in Milwaukee to pressure its city council to produce a few landscape parks out of large tracts of land they donated to the city (Gurda, 2006). It was hoped these environmental amenities would quell labor unrest. A few parks were not enough, however, to compensate for the squalor and unsanitary conditions in which most of Milwaukee's proletariat was living. The working class in Milwaukee subsequently elected "Sewer Socialists" to many positions in municipal government during the first half of the 20th century, including the mayor's office (Beck, 1982; Tolan, 2003). The socialists, unlike their liberal predecessors, taxed capital in the city to build collective urban environmental infrastructure including sanitary sewers and a large public park system for the benefit of workers and their families (Anderson, 1987). Charles Whitnall was a socialist politician who is considered the father of Milwaukee County Parks.[1] He said in this regard, "Milwaukee's principle asset is its mechanical manpower. We have more highly skilled mechanics than any other city in the country … To preserve it and develop it further, these craftsmen must move where their children can be close to mother earth and in touch with the natural environment" (cited in Tolan 2003: 21). Socialism lived up to its promise to enhance the quality of Milwaukee's environment but as a radical political platform it quickly fell out of favor at the beginning of the Cold War. Keynesian-oriented municipal officials that replaced the socialists had the consent of capital and the urban citizenry to continue to build the park system to its present size of nearly 15,000 acres by 1980. They did so by spending nearly one-third of the County tax levy on parks they believed enhanced the standard of living for residents and attracted more business investments. However, the longstanding "negotiation" between the municipal government, capital, and civil society regarding Milwaukee's parks seemingly broke down in the early 1980s.

The Milwaukee County Parks board was disbanded in 1982 and a series of fiscally conservative County Executives since then used their direct control over the Parks Department to drastically cut parks funding to balance the County budget. Adjusted for inflation, the 2010 budget for parks stands at less than half of what it was in 1980. A 2009 audit of the park system conducted by the Milwaukee Department of Audit reports the park system is really now a "Tale of Two Systems" where a few highly visible parks downtown are spectacular, while many others in lower profile parts of the County are in need of 270 million dollars in repairs and are being closed (Greater Milwaukee Committee, 2008; Heer and Jenkins, 2009). This bifurcation represents a renegotiation between the municipal government, capital, and civil society concerning the role parks play in shaping urban social relations. At stake in the renegotiation is the former idea that a high standard of urban living is enhanced through the collective production and consumption of public environmental amenities like parks systems. Today, neoliberal modes of environmental governance rearticulate this relationship by linking the production and consumption of specific parks to civil sector partnerships that prioritize market profitability and personal responsibility. The remainder of this chapter therefore demonstrates increased urban vulnerability is a result of this process when: 1) parks systems are disinvested by the local state so capital interests can lease specific park spaces most useful for selling expensive commodities; 2) the

---

1   Whitnall and the Milwaukee County Parks Board acquired most of the City of Milwaukee Parks and transferred them to the Milwaukee County Parks Department during the Great Depression. I use "urban" and "county" parks interchangeably in this paper to refer to the Milwaukee County Parks System as nearly all of Milwaukee County is urbanized.

construction of civil sector partnerships through volunteering in specific parks "justifies" the reduction in unionized municipal workers who used to care for all of the park spaces in the system; and finally, 3) these factors make poor and minority citizens susceptible to a diminished standard of living when their opportunities to recreate and rejuvenate in safe park spaces become dependent on their ability to participate in these market practices.

Civil sector partnerships for parks provision and the production of vulnerability in Milwaukee have not occurred in a political economic vacuum. Geographers writing about neoliberalization in the West documented the dismantling of the Keynesian Welfare State and its interventionist economic, social, and environmental policies following the crisis of Fordism in the 1970s and early 1980s (see, e.g., Peck and Tickell, 1992). They also recognized the devolution of economic regulatory capacity away from the scale of nation state toward localized urban regions (Jessop, 1997). One result is the development of a global marketplace where increasingly competitive cities race for footloose capital investments (Peck, 2001). Municipal governments have accordingly undertaken initiatives to help unleash the maximum potential and profitability of their own capital markets in part by transferring resources from their social service programs and physical infrastructures into activities that benefit the private sector (Lemke, 2001; Jessop, 2002; Swyngedouw, 2005).

For example, assaults on municipal labor unions reduced the payroll costs of public services and in many instances opened up new market opportunities for private employers to fill the employment void (Brenner and Theodore, 2002). Municipal contracts with corporations make many kinds of formerly public infrastructures less costly to local government and more profitable to the private sector. Municipal water systems are increasingly leased or sold to private firms that turn water into a commodity sold to customers at a profit (Bakker, 2005; Prudham, 2007; Loftus and Lumsden, 2008). Cities lease parks to firms to generate land rents while the lessees earn profits through the sale of associated commodities and services on parklands (Perkins, 2009a, 2009b). Urban forests are increasingly planted and maintained in highly visible parts of the city that generate wealth, including the central business district and wealthy neighborhoods- while poorer neighborhoods continue to lose trees (Heynen and Perkins, 2005).These kinds of initiatives helped turn cities into entrepreneurial spaces that offer ever-larger public subsidies to facilitate increased private capital investment in (re) development projects and other public/private ventures (Brenner et al., 2002; Ward, 2007). Many of these studies suggest neoliberalization carried forth through the transfer of public resources to the private sector benefits wealthy people while making the poor increasingly vulnerable as urban environmental services are put out of their reach. The next sections of this chapter elaborate how this process plays out in Milwaukee's urban parks.

Fiscal Austerity, Disinvestment, and Market Opportunity in the Parks

Budgets for public parks and many other social and environmental services in Milwaukee have been systematically cut for nearly two decades. Scott Walker, the Milwaukee County Executive from 2002–2010, was particularly adept at cutting funding for public services and infrastructures- especially the park system.[2] Walker also had a record in County government for reducing taxation. This combined with losses in shared revenue from the State of Wisconsin means County coffers have not even been able to keep up with the rate of inflation for nearly eight years (Henken, et al., 2009). Walker consistently vetoed any attempts by his Board of Supervisors to raise revenue for County services and infrastructures, including public parks, through increased taxation. Milwaukee County now has an 80,000,000 dollar structural deficit that could legally be made up within the County by marginally raising its property tax rate. However, it is widely acknowledged property taxes will not be raised to make up the deficit (Henken and Allen, 2010). Thus it is quite unlikely that the 270,000,000 dollars needed for infrastructural repairs in the parks will be made up through taxation anytime soon, either.

Parks disinvestment is associated with Walker's conscientious maneuver to redefine collective investment in infrastructural assets as a fiscal *liability* to private property owners. This is a significant break from past ideas about the benefits collective parks investment provide to the owners of private property in Milwaukee. Historically speaking, many taxpayers believed green space acquired and maintained through collective investment provided everyone a higher standard of living, in turn benefitting the urban economy by creating healthier, more productive workers. It was also understood in Milwaukee that well-maintained parks bolstered the market value of adjacent private real-estate properties. This positive logic regarding the economic benefits of parks investment has been jettisoned, however.

Scott Walker is instead on record in his 2010 Budget Address as saying that public services and infrastructures, including the parks, are too expensive and therefore threaten the solvency of the County (Henken and Allen, 2010). He also stated he was unwilling to skirt insolvency by raising taxes because he believes taxes unduly burden businesses (read capital) and home owners. He went on to say in his address, "… for the eighth year in a row, I am presenting a County budget that does not raise the property tax levy from the previous year. In addition to holding property taxpayers harmless, this budget does not rely on increased sales taxes or implementation of a new wheel tax to fund County government. Increasing these taxes places an added burden on struggling families and employers …" Walker was literally telling Milwaukeeans that instead of enhancing their standard of living, parks and other public services *threaten* it if their cost is allowed to drain

---

2   Scott Walker was elected Governor of Wisconsin and therefore left his County Executive Position in 2011.

the local economy of its productive capacity and increase property owners' risk of tax delinquency.

Executive Walker's rhetoric of insolvency and his actions to disinvest in the park system provided him with what he believed was an opportunity to turn some of Milwaukee's public parks directly into private property. The dilapidated condition of many parks, a lack of active programming, and increased user fees have translated into decreased patronage at many parks throughout Milwaukee (Heer and Jenkins, 2009). Thus Walker justified the idea of selling hundreds of acres of parks, particularly in the southern part of Milwaukee County, to real estate developers to reduce costs and help cover budget deficits. Ultimately he was unsuccessful in doing so as his proposal was not passed by the Milwaukee County Board of Supervisors. But he has successfully subjected the parks to the vagaries of the real estate market without directly selling them, however.

Walker used the poor condition of the public parks combined with their relative location to attract private vendors into some of them. There are striking parallels here to Smith's rent gap theory concerning private property and gentrification in older sections of cities (1996). Smith suggests private and/or tax forfeited property is devalued on the real estate market to the point where it can be purchased by developers for very low prices. Relative location, historic building structures, among other considerations help developers determine when and where they should purchase disinvested properties, reinvest and renovate them, and sell them for a hefty profit on the market. It is also known as putting a property to its "highest and best use" when the market timing and geography is just right. Obviously disinvested public parks are different than tracts of privately owned, disinvested housing. However, some parks in Milwaukee – particularly those located downtown and along the lakefront-have unique structures on site and would likely be worth millions of dollars on the private real estate market if sold to commercial developers. The "rent gap" here exists in the sense that parks near the central business district and the lakefront usually generate little to no revenue to cover the cost of their operation to the municipality, yet they have the potential to generate substantial revenue through land rents and private commodity sales.

The Parks Department, under the direction of Walker, used these "rent gap" circumstances to pursue contracts with private capital in some of its high profile parks (Perkins, 2009a). Building and exploiting private market capacity into collective infrastructural resources is precisely what makes this process *neo*liberal.[3] The Parks Department chooses park facilities it feels will support profitable vendors and by extension generate land rents for the county. A bidding process is subsequently initiated whereby private companies submit proposals to operate the site for the sale of their commodities. Vendors set up businesses in a number of highly visible parks; but we are not talking about popsicle stands for the kids. Instead vendors transform historic park buildings into bistros, coffee shops, and warming houses that sell expensive foods and beverages. Alterra

---

3   See Perkins, 2009a.

Coffee at Veteran's Park, Starbucks Coffee at Red Arrow Park, and Bartolotta Restaurant Group's Lake Park Bistro and its concession stand at the Bradford Beach Boathouse are just a few of these public/private partnerships based on high-end commodity consumption. Walker is very much in favor of creating more of these partnerships. He said, "If it takes a coffee shop in every park to keep them open, then I'm all for it" (Umhoefer, 2006). The Milwaukee County Department of Audit and the supposedly nonpartisan think tank, Public Policy Forum suggest the Parks Department aggressively pursue more of these partnerships (Heer and Jenkins, 2009; Henken et al., 2009), despite evidence the land rents these businesses pay the County are low compared to what could be charged based on their locations (Perkins, 2009a).

Partnerships with private capital might be a harder sell to the public if the parks were not already in poor condition and lacking patronage. Instead this is a situation where few people are openly concerned that statist forms of intentional disinvestment in their parks create spaces of market opportunity exploited by private capital. Here is precisely where urban vulnerability is quietly produced in relation to the renegotiation of the provision and purpose of formerly public parks. These commodity parks are no longer invested for the sake of equitably providing everyone a high standard of living through recreation and rejuvenation in green space. Rather they are privately invested and programmed to (re) produce market relations through patronage based on the consumption of high end commodities. People who can afford to patronize commodity parks will find these spaces continue to fulfill their legacy of providing places to recreate and rejuvenate. Those who cannot afford to patronize them potentially lose the benefits those spaces were originally designed to provide. Thus their standard of living is put at risk as their opportunities to recreate and rejuvenate are diminished. Even if the cost of patronage in these commodity parks was somehow not an issue-this is a highly uneven strategy for parks renovation. Dozens of central city parks have yet to attract any private capital whatsoever; thus residents there find their opportunities to safely recreate and rejuvenate spatially circumscribed (Heer and Jenkins, 2009).

Replacing Paid Parks Employees with Civil Sector Partnerships?

Scott Walker was elected to the position of County Executive in 2002 after his predecessor, Executive Ament, resigned amid a recall campaign against him for signing off on a generous pension plan for County employees (Umhoefer et al., 2009). Walker proclaimed during his election campaign that the cost of paying people to maintain County services, including parks, was out of control. He accordingly fulfilled his campaign promise to reduce the size and cost of County government by reducing the number of workers on the municipal payroll. The Director of the American Federation of State, County, and Municipal Employees (AFSCME) 48 was interviewed for this project regarding the personnel cuts Walker made to the County workforce.[4] He stated plainly that under Walker's tenure, 1500 positions had been eliminated County-wide of which 600 were direct layoffs. The President of AFSCME Local 882 that specifically represents park workers was also asked what Walker's downsizing policies meant to his workforce. He replied during the interview:

> Well, it's been destroyed. It's next to nothing now ... The lower level park maintenance workers are all gone. They have been laid off- they do the bulk of the manual work. Now since our seasonal budget was slashed by 40 percent, we have only low level seasonal employees that can't do as much work because they are limited in scope by their licenses, etc. So the people that are left are limited to what duties they can do, and we only have so many positions in the specialized areas that have CDLs [commercial drivers licenses] ... all told I bet we are only running fifty full time positions.

Fifty full-time positions is a small fraction of the 1300 full-time employees working in the parks in 1980. These interviews with the AFSCME representatives came on the heels of more Walker efforts to downsize the entire County workforce.

In early 2010 Walker laid off another 76 public employees, including 25 park maintenance workers (Kaiser, 2010). However, these layoffs pale in comparison to the concessions he foisted on his workforce in his 2010 budget. In 2009 Executive Walker's negotiating team hammered out a labor contract with AFSCME 48 that would have secured no layoffs for two years in exchange for no employee raises or enhancements in their benefits. However, Walker disregarded that tentative contract and created a 2010 County budget that sought an additional 41 million dollars in concessions from his employees. His 2010 budget included the elimination of 400 more jobs, privatization of 222 additional positions, a 3 percent wage decrease for all remaining employees, increased health care premiums, and 22 furlough days. He defended his Draconian measures in his 2010 Budget Address when he said:

> Today over 48 percent of the county budget goes to fund wages and benefits for County employees, and the cost of benefits is growing at an alarming rate ... With so many private sector workers seeing their wages and benefits frozen or cut to preserve jobs, it is hard *not* to expect the same from those in government ... we must negotiate additional concessions to help balance the escalating cost of public sector employee benefits for taxpayers, now and in the future. Without further action, Milwaukee County will become insolvent. (emphasis in original)

---

4   All of the interviewees quoted in this chapter were interviewed as part of a larger, as of yet unpublished, project on shared forms of urban green space governance within neoliberal political economy.

AFSCME 48 formally charged Executive Walker and the Milwaukee County Board of Supervisors with bad faith bargaining under Wisconsin collective labor law. It remains to be seen how this charge will play out and how this will affect parks workers specifically.

The Milwaukee County Parks Director was also interviewed to see how she is coping with the loss of support staff in her parks. She is in a difficult position because her boss, Executive Walker, has consistently cut her annual budget during her seven years as Parks Director. She stated, when asked about numerous employee layoffs, that she actually prefers "a stealth workforce in the parks." It was evident in the interview that she does not push her boss for more paid employees and is not planning for increased numbers of paid employees as a long-term strategy for parks provision. Instead she, like the former Executive, is planning neoliberalized, market-based solutions to compensate for fewer paid workers.

She is quite busy planning partnerships with local businesses that provide volunteers to work in the parks. Companies like Briggs & Stratton, Harley-Davidson, and Diversified Insurance have teamed up with the Parks Department to clean up green spaces. She got a local architectural firm to donate its time and expertise to draw plans for renovating McKinley Park. She teamed up with Johnson Controls to form a Youth Conservation Corps that gets urban kids volunteering to work in the parks. She solicits donations from private philanthropists to revamp park structures like the Boathouse at Bradford Beach on Lake Michigan. The Parks Director also has been negotiating with the nonprofit Urban Ecology Center in the hopes the organization will adopt ten parks and set up its educational outreach centers in each one. She admits she is at the head of the curve when it comes to generating community capacity for parks partnerships. She was quoted in the *Milwaukee Journal Sentinel* as saying, "We're in the bridge age. We don't know where we're going to land with this. You're getting to the next shore" (Glauber, 2010). The Parks Director has been so aggressive and successful in pursuing community partnerships that she was awarded the "2009 National Recreation and Park Association Gold Medal Award for Excellence in the Field of Park and Recreation Management." It is important to note that the award is for *parks management*. But former Executive Walker incorrectly boasted Milwaukee County won the award for the best *park system* in the U.S. on his website and in the media (Walker, 2009).

The Parks Director also works with nonprofit organizations to generate other forms of citizen volunteerism. For example, the nonprofit, Park People of Milwaukee County (PPMC) acts as a liaison between the Parks Department and the more than 60 "friends of the parks" groups that have arisen to care for disinvested neighborhood parks. PPMC views its role in part as building volunteer capacity in the friends groups to compensate for diminished municipal budgets for urban green spaces. The nonprofit also coordinates its own volunteer cleanups and "weedouts" where urbanites can activate their environmental citizenship by working in parks to remove invasive species like garlic mustard and buckthorn. The PPMC Director interviewed for this project stated that his volunteers are frustrated they cannot do more work in the parks besides the cleanups and weedouts. However, he acknowledged that PPMC and its volunteers are bound to a contract with AFSCME 48 Local 882 signed in 1985 that limits their activities to tasks not described in labor contracts negotiated with the County. In other words, AFSCME 48 long ago anticipated volunteers might be used by the County Executive and Parks Director as low-cost labor replacements for unionized workers.

The interviewees from AFSCME 48 and PPMC all acknowledge that environmentally–oriented citizen volunteers in the parks are not going to make up for laid-off paid laborers anytime soon. The magnitude of need for general maintenance greatly outweighs the current number of volunteer hours logged in the parks every year. Many volunteers are also not trained or qualified to perform complicated tasks that necessitate working with heavy mechanical equipment. However, the Director of AFSCME 48 stated that in light of Walker's budget cuts and the Park Director's numerous community-based initiatives "there is a lot more paranoia" in his workforce concerning citizen volunteering. In other words, his workers recognize their vulnerability to these developments that threaten the long-term stability of their positions. Therefore union workers and citizen volunteers will continue to work uneasily alongside each other in the parks into the foreseeable future.

Conclusion: Urban Vulnerability and Market-oriented Parks Governance

Past forms of park provision have never been perfect. In fact the social engineering projects behind collective parks provision in the past remind us that many urban green spaces were built for the purpose of reproducing stable class relations in cities. Milwaukee is certainly no exception. Despite this rather insidious intent, collectively produced green spaces provided opportunities for the working and middle classes in Milwaukee to recreate and rejuvenate, therefore enhancing their standard of living. Municipal government, capital, and workers alike decided green space was therefore a necessary, and ultimately profitable, public investment. County parks also long served the purpose of providing contact with nature to poor people who otherwise would not be able to do so outside Milwaukee. The Director of AFSCME 48 said during his interview:

> I didn't even get into my whole thinking about how what Scott Walker does is a fundamental attack on poor people in our community. People of color and poor people in our community. But I do believe that those of us with the means have the opportunity to go play golf at a private golf club. We have the means to go and recreate at places that frankly people without those means don't have access. And the parks are essential in our communities in order to provide the venues for folks who lack other resources to recreate.

This concern demonstrates why it is important to consider how urban vulnerability is necessarily produced when the provision for parks is renegotiated between municipal government, capital, and the civil sector in neoliberal, profit-oriented terms.

Disinvestment in the parks during the last 30 years has reduced patronage as equipment malfunctioned, people began to perceive the parks as increasingly dangerous places to visit, and facilities closed (Heer and Jenkins, 2009). Former Executive Walker's record suggests he does not believe disinvestment in, and decreased patronage of, parks makes poor and working people in Milwaukee vulnerable to a reduced standard of living when their opportunities to recreate and rejuvenate are diminished. However, Walker *was* explicitly concerned about what he sees as another kind of vulnerability when he stated in his 2010 Budget Address, "Increasing taxes places an added burden on struggling families and employers ..." In other words he describes continued public investment in parks and other services as an activity that makes business owners and home owners increasingly vulnerable to over-taxation. The solution to disinvestment in the parks system according to him is to bring them within the purview of the real estate and commodities markets- in turn making them less costly to the taxpaying public and more profitable to private capital.

The result is the Park System gets some rent while the vendors make large profits selling their commodities (Perkins, 2009a). This process creates an internally-differentiated park system whereby the equity built into the system is undermined. Here is where we need to critique Walker's solution to his version of urban vulnerability. Office workers and well-to-do residents with cars can afford to patronize lakeside parks with elite vendors and therefore reap the benefits of the new market provision model. In theory they are free to spend more at these vendors because they pay fewer taxes toward the maintenance of the entire park system and other County services and infrastructures. The parties that potentially lose the most in this marketized renegotiation of green space provision are the poor and working class people that the AFSCME Director expressed concerned about in his interview. Coffee shops and bistros are not adopting parks in neighborhoods away from the lakefront and downtown. Those parks without the potential to generate land rent through commodity sales continue to be disinvested as the Parks Department markets parks in more advantageous locations. Intuitively, it seems then that marketized parks provision constitutes an environmental injustice.

It also goes without saying that former Executive Walker's war against his County employees makes them increasingly vulnerable as well. Wage cuts, furlough days, increases in healthcare premiums, and other benefits reductions all jeopardize his workers' ability to provide for themselves and their families. Workers still on the payroll might consider themselves fortunate, however. Walker laid-off nearly 600 workers during his tenure; many of those employees worked for the Parks Department. Those workers lost jobs that provided relatively decent wages, though the Director of AFSCME 48 said the average County worker makes approximately 32,000 dollars per year. The President of AFSCME 48 Local 882 identified a central contradiction in Walker's production of vulnerability in public sector employment. He said in his interview, "If you are a County employee in Milwaukee County, you have to reside in the County. Which means of course your taxes, your mortgage, your rent, your consumer goods are generally purchased in the County. It's like an ecosystem." In other words he was saying that in addition to performing a public service through their jobs, paid public sector employees generate more economic activity in Milwaukee. He went on to state Walker's layoffs actually hurt the local and state economy because most unemployed County workers are "trying to get on Badger Care [subsidized health care] and unemployment … that is being of course paid for by the taxpayers in the County and throughout Wisconsin." His point is well-taken. These paid parks positions allowed individuals to shelter, feed, and clothe their families while providing necessary environmental services to everyone else in the city. Now they are not providing that valuable service while they instead seek welfare assistance to meet their daily needs. The important lesson from Walker's actions then is that everyday life becomes a more brutal, individualized calculus for both unemployed County workers and those people who depended on them to provide healthy green spaces.

But what about community volunteers in the parks? Are they not working to make up for lost paid parks employees in those parks not leased by capital? And if they volunteer are they not at the very least mitigating some of the vulnerability people experience when they lack quality green spaces in which to recreate and rejuvenate under marketized provision? The answer to these questions is a tentative, "yes." However, in a climate of perpetual paid labor force reductions, shared environmental governance through volunteering is not adequate to meet the needs of the park system as a whole. Those workers lost to layoffs worked throughout the entire park system, on all parks, in all locations. Citizen volunteers coordinated by PPMC work in specific parks to do cleanups and weedouts, but are unable to do much more. Those volunteers for PPMC are usually white people with educations and decent incomes, according to the PPMC Director. Thus minority groups in the city are largely left out of these efforts. Organized "friends of the parks groups," by extension, are usually comprised of citizens who live in wealthy communities and want to work on local parks in their own neighborhood (Perkins, 2010). They are frequently limited to cleanups and fundraisers to fix buildings or equipment in a specific park. In some cases they are successful, but other parks in the central city lack well-to-do citizen groups that can adequately compensate for systemic, municipal disinvestment in their local green spaces.

Volunteer capacity building for parks governance ultimately creates vulnerability because the production of a high standard of living through contact with urban green space is no longer a collective responsibility, but is dramatically individualized. This happens because the renegotiation of a parks compact under neoliberalization requires the activation of urban citizenship through individuals' ability to participate in various forms of volunteer activity. We need to further investigate the ramifications of this push to build citizen capacity in neoliberalized parks governance. Why, e.g., are minorities and poor people not better represented in the ranks of parks volunteers though they comprise nearly half Milwaukee's population? Why are there fewer "friends of the parks" groups in poor neighborhoods? At any rate, the answers to these questions will demonstrate what we already know – individualized, citizenship-oriented initiatives are proving to be too little, too late to mitigate the magnitude of park system disinvestment.

It is hard to be optimistic in the wake of the former County Executive's severe budget cuts for parks and his supposed market solutions. It is even more difficult with the realization that the crisis in the system has been building long before Walker's tenure and will very likely continue to deepen even after he abdicated his post for the Wisconsin Governorship. This tells us the sea change of the last thirty years regarding the provision for urban green space is much larger than one, austere man in a powerful position to make fiscal policy. Priorities by many people, particularly in Milwaukee's fiscally conservative suburbs, have likely changed regarding the collective production and purpose of parks- and by extension- their deployment as collective prevention of urban vulnerability. Whereas one-third of the County budget was designated for parks in 1980; proponents have been unsuccessful in generating even the smallest dedicated tax increase to offset a 60 percent reduction in the parks budget since then.

References

Anderson, H. 1987. Recreation, entertainment, and open space: park traditions in Milwaukee County. In: R. Aderman (ed.) *Trading Post to Metropolis: Milwaukee County's First 150 Years*. Milwaukee: Milwaukee County Historical Society.

Bakker, K. 2005. Neoliberalizing nature? market environmentalism in water supply in England and Wales. *Annals of the Association of American Geographers* 95(3) 542–565.

Beck, E. 1982. *The Sewer Socialists: A History of the Socialist Party of Wisconsin 1897–1940, Two Volumes*. Fennimore: Westburg Associates.

Brenner, N. and Theodore, N. 2002. Cities and geographies of "actually existing neoliberalism". In: N. Brenner and N. Theodore (eds) *Spaces of Neoliberalism: Urban Restructuring in North America and Western Europe*. Oxford: Blackwell, 2–32.

Brenner, N., Jessop, B., Jones, M. and MacLeod, G. 2002. *State/Space: A Reader*. Oxford: Blackwell.

Cranz, G. 1989. *The Politics of Park Design: A History of Urban Parks in America*. Cambridge: MIT Press.

Glauber, B. 2010. Sue Black stays on a relentless quest for improvements despite lack of cash. May 9, 2010 Edition of the *Milwaukee Journal Sentinel*. <http://www.jsonline.com/news/milwaukee/93248069.html>. Last accessed 8-16-2010.

Greater Milwaukee Committee 2008. *Cultural Asset Inventory of the Milwaukee 7 Region*. Milwaukee: Greater Milwaukee Committee Publication.

Gurda, J. 2006. *The Making of Milwaukee*, 3rd Edition. Brookfield: Milwaukee County Historical Society.

Heer, J. and Jenkins, D. 2009. *A Tale of Two Systems: Three Decades of Declining Resources Leave Milwaukee County Parks Reflecting the Best and Worst of Times*. Milwaukee: Milwaukee County Department of Audit.

Henken, R. and Allen, V. 2010. *Budget Preview: 2011 Milwaukee County Budget*. Milwaukee: Public Policy Forum.

Henken, R., Allen, V. and Dickman, A. 2009. *Budget Brief: Milwaukee County 2010 Executive Budget*. Milwaukee: Public Policy Forum.

Heynen, N. and Perkins, H. 2005. Scalar dialectics in green: urban private property and the contradictions of the neoliberalization of nature. *Capitalism Nature Socialism* 16(1), 99–113.

Jessop, B. 1997. A neo-gramscian approach to the regulation of urban regimes: accumulation strategies, hegemonic projects, and governance In: M. Lauria (ed.) *Reconstructing Urban Regime Theory, Regulating Urban Politics in a Global Economy*. Thousand Oaks: Sage, 51–73.

Jessop, B. 2002. Liberalism, neo-liberalism and urban governance: a state theoretical perspective. *Antipode* 34(3), 452–472.

Jones, K. and Wills, J. 2005. *The Invention of the Park: From the Garden of Eden to Disney's Magic Kingdom*. Cambridge: Polity.

Lemke, T. 2001. The birth of bio-politics: Michael Foucault's lectures at the College de France on neo-liberal governmentality. *Economy and Society* 30(2), 190–207.

Loftus, A. and Lumsden, F. 2008. Reworking hegemony in the urban waterscape. *Transactions of the Institute of British Geographers* 33, 109–126.

Low, S., Taplin, D. and Scheld, S. 2005. *Rethinking Urban Parks: Public Space and Cultural Diversity*. Austin: University Texas Press.

Peck, J. 2001. Neoliberalizing states: thin policies/hard outcomes. *Progress in Human Geography* 25(3), 445–455.

Peck, J. 2001. *Workfare States*. New York: The Guilford Press.

Peck, J. and Tickell, A. 1992. Accumulation, regulation and the geographies of post-fordism: missing links in regulationist research. *Progress in Human Geography* 16(2), 190–218.

Perkins, H.A. 2009a. Turning feral spaces into trendy places: a coffee house in every park? *Environment and Planning A* 41, 2615–2632.

Perkins, H.A. 2009b. Out from the (green) shadow: neoliberal hegemony through the market logic of shared urban environmental governance. *Political Geography* 28(7), 395–405.

Perkins, H.A. 2010. Greenspaces of self-interest within shared urban governance. *Geography Compass* 4(3), 255–268.

Prudham, S. 2007. Poisoning the well: neoliberalism and the contamination of municipal water in Walkterton, Ontario. In: N. Heynen, J. McCarthy, S Prudham, and P. Robbins (eds) *Neoliberal Environments: False Promises and Unnatural Consequences*. Routledge: London, 163–176.

Rosenzweig, R. 1983. *Eight Hours for What we Will, Workers and Leisure in an Industrial City, 1870–1920*. Cambridge: Cambridge University Press.

Rosenzweig, R. and Blackmar, E. 1992. *The Park and the People: A History of Central Park*. Ithaca: Cornell University Press.

Schultze, S. 2009. Walker seeks across-the-board wage cuts in budget. September 24, 2010 edition of the *Milwaukee Journal Sentinel* newspaper.

Swyngedouw, E. 2005. Governance innovation and the citizen: the janus face of governance-beyond-the-state. *Urban Studies* 42, 1991–2006.

Tolan, T. 2003. *Riverwest: A Community History*. Milwaukee: Past Press.

Umhoefer, D. 2006. Walker looks to coffee shops to perk up parks: county executive talks of dire budget year ahead. *Milwaukee Journal Sentinel*, 6 February.

Umhoefer, D., Schultze, S. and Diedrich, J. 2009. Milwaukee County pension scandal trial primer. May 2, 2009 edition of the *Milwaukee Journal Sentinel*.

Walker, S. 2009. County parks win national gold medal award for excellence October 14, 2009. <http://www.milwaukee.gov/ImageLibrary/Groups/cntyExecutive/2010_Proposed_Budget__County_Parks_Win_National_Gold_Meda_.pdf> Last accessed: 9-27-2010.

Ward, K. 2007. Creating a personality for downtown: business improvement districts in Milwaukee. *Urban Geography* 28(8), 781–808.

Part 2

Unanticipated Vulnerabilities: Sustainability Planning, Environmental Movements, and Activism

Chapter 5

Re-imagining the Local: Scale, Race, Culture and the Production of Food Vulnerabilities

Julian Agyeman and Benjamin L. Simons

Introduction

Over the past decade, the production and consumption of locally grown food has been accepted as the ultimate panacea for many alternative food movement advocates, with local food systems emerging as an almost synonym for urban sustainability. For many "locavores," local food systems provide the antidote to the destructive tendencies of global capitalist *industrial* agriculture by offering ecologically sound agricultural practices; support for small-scale family farms and local economic development; fresher and healthier food for the consumer; greater democracy and transparency in food systems decision-making; and a more holistic connection between the consumer, the farmer, and the rural landscape. The local food movement has garnered significant attention in the mainstream media and participation among urban residents as consumers and producers (see Pollan, 2006 and Kingsolver, 2007). Rising interest has been matched with a proliferation of farmers' markets and Community Supported Agriculture (CSA) projects, as well as with increasing numbers of food co-ops, groceries, restaurants, and food services operations of large institutions across the country sourcing more of their food locally. Local food boosters have framed their efforts by drawing on a variety of perspectives ranging from rural development, to sustainability/ environmentalism, to communitarianism and anti-globalization.

What is lacking, however, in most of the locavore discourse is an explicit recognition of equity and social justice concerns in terms of the ability of economically marginalized and minority urban populations to produce, access and consume healthy and culturally appropriate foods, and, additionally, how the development of local food systems relates to these dynamics and power asymmetries (see Allen, 2008, 2010). Due to the wave of neo-liberal economic and political forces that increasingly shape the development of the urban environment, communities that have traditionally held economic and political power have been provided the resources, infrastructure and investment to allow them abundant access to and control over food sources, while those communities lacking significant economic resources and political power (typically low income communities and communities of color) have been ignored, under-resourced and disinvested thereby deepening their lack access to, and control over food sources (see Heynen, 2006). There are several implications to this food insecurity and the associated "food deserts" (Wrigley, 2002) in low income/people of color neighborhoods, including nutritionally poor diets that reinforce and compound well documented public health crises related to hypertension, obesity and diabetes in these areas (see Diex Roux et al., 2006 and Giang et al., 2008).

For the purposes of this chapter, we develop and apply the concept of food vulnerability to highlight the interactions among food production, food access and political and economic asymmetries, with particular analytic focus on the ways in which these interactions render already vulnerable populations increasingly vulnerable. In this chapter we will demonstrate the unanticipated and conventionally unrecognized consequences of local food movements for vulnerable populations in the urban context.

In addition to improving processes of food production and consumption, the local food movement has also been espoused to facilitate the creation of both ecologically and socially just urban environments. However, we argue that patterns of local food system development so far mirror, in many ways, those of the conventional food system. Of particular note has been the establishment of unjust spatial access to locally produced, healthy and affordable food within urban regions. Farmers markets, CSA drop-offs, restaurants with locally sourced menus, and local food vendors are often concentrated in middle to upper income neighborhoods while lower income, predominantly minority communities often lack similar access (see Allen et al., 2006; Allen and Hinrichs, 2008 and Hinrichs, 2003). Vulnerability, in this situation, refers to the disproportionate risk of poor nutrition and food insecurity experienced by low-income minority communities. The mainstream local food movement described by Pollan (2006) and documented in periodicals such as *Edible* largely ignores communities who are made most vulnerable to poor nutrition and food insecurity through processes of uneven development resulting from histories of under- and disinvestment practices. This, in conjunction with current day conditions of reduced access, widen food access divide. Food vulnerability can therefore be considered both as condition of having reduced access to locally grown, healthy food and as process driven and as a process that is perpetuated by historical practices of urban land development and economics.

There is an emerging discourse that addresses the vulnerabilities associated with the local food movements. In the same way as "environmental justice" arose out of a disquiet over control of the environmental agenda in the 1980s, and "climate justice" is growing as a result of a lack of equity and justice considerations in the mostly science-economics focused climate campaigns, discourses around "food justice," "food democracy" and "community food security" are rising to the top of the agenda for many food system scholars and activists. Increasingly, politically engaged local food initiatives, such as urban agriculture, community gardening, farmers markets, CSAs and grocery co-ops are developing as a means to reverse the trends of poor nutrition, health and food vulnerabilities in underserved urban communities (Baker, 2004). Many non-profits have emerged in cities across the country in recent years to address these issues, with some to great effect (see Community Food Security Coalition, Growing Power, and the Food Project). A growing body of scholarship is critically analyzing the discourse and ability of conventional local food systems to address issues about how food vulnerabilities are both produced and experienced, including insensitivity to cultural difference of diets, reduced access to farmers markets and cooperatives, and disproportionate exposure to unhealthy foods that are concentrated in low-income communities of color. With its emphasis on the ecological benefits associated with local organic food production and consumption, literature on local food movements fail to engage one perhaps unintended yet significant consequence: culturally specific diets. Many of the factors involved in producing food vulnerability for low-income communities of color by conventional food systems are acting upon and within these food systems, thus reinforcing disparities. Ultimately, this process of reinforcement undermines the promise of food local systems as a resilient strategy. We argue that small (mostly white) famer systems of production for healthy yet culturally exclusive diets contribute to the deepening of the inequities coming out of conventional local food systems. In doing so, issues of equity and social justice are not engaged, are not made evident, and are therefore at risk of perpetuating disparities.

For all the potential that exists in moving away from globalized capitalist industrial agriculture towards more localized, small-scaled production and consumption of food, a new set of conflicts and contradictions are generated when framing the "local" as more sustainable. This is especially true when local food systems produce a two-tier system of inclusion and exclusion, and when the project of localization itself is imposed upon low-income urban areas as opposed to being generated and determined by the residents of those areas themselves. This chapter examines the conflicts and contradictions of local food systems through the lens of the production of food vulnerabilities, with the goal of forming a more robust understanding of the true potential of local urban food systems to create food secure, healthy, and resilient communities.

Understanding Urban Food Vulnerabilities through an Urban Political Ecology Lens

Recent academic and activist attention has addressed the production of food vulnerabilities, especially the processes responsible for the formation of food deserts. Food deserts are the result of a history of disinvestment in and neglect of mostly low-income urban and rural areas, which have not been recognized as profitable sites for supermarket and grocery store location and have therefore been left with limited and often less healthy, more expensive options for food access, such as corner stores and fast food establishments. While food vulnerabilities are most pronounced in food deserts, not all those suffering food vulnerabilities live in food deserts. When situated within an urban political framework, the production of food vulnerabilities is driven by asymmetrical power relationships involved in the transformations of the urban environment. Heynen and Swyngedouw note that, "the aim [of urban political ecology] is to expose the processes that bring about highly uneven urban environments" (2003, p. 906). Here the built environment and urban processes are not understood as separate from nature and natural ecological processes. Instead, they argue that the constant evolution of cities is, in Marx's terms, a metabolic process of humans transforming "nature" into the built environment, creating the socio-ecological landscapes and socio-cultural conditions that comprise the city. Moreover, these "environmental transformations are not independent from class, gender, ethnicity, or other power struggles and, in fact, often tend to be explained by these social struggles," and that "these metabolisms produce a series of enabling (for powerful individuals and groups) and disabling (for marginalized individuals and groups) social and environmental conditions" (2003, p. 911).

In capitalist, market driven societies, the ways in which these metabolic processes and environmental transformations occur are primarily dictated by those who control the wealth and power as "it is this nexus of power and the social actors carrying it that ultimately decide who will have access to or control over and who will be excluded from access to or control over resources and other components of the environment" (Heynen and Swyngedouw, 2003, p. 911). The result is spatial injustice with respect to the development of the urban of environment and distribution of resources and amenities within cities. Those with less wealth and political power consequentially suffer from the uneven development and disproportionately experience the negative effects of urbanization.

Food is one of the most basic and fundamental environmental resources that humans metabolize, as "human bodies are produced through socio-metabolic processes that link their existence to external processes that produce food," but forces beyond individual and community control largely determine who has access to what types of food and where (Heynen, 2006, p. 129). In societies where food is viewed not as a fundamental human right, but as a commodity to be bought and sold, access to food is dictated by one's economic ability to purchase food, and the pricing of food is based on costs associated with its production and profit margins. Since the vast majority of people don't grow their own food, especially in urban areas, and rely on food retailers and food services for access, these food outlets and those that supply them are the primary source of food. Supermarket and grocery chains have determined over the past half-century that it is more profitable to locate stores in middle to upper class urban and suburban neighborhoods and have abandoned many lower income urban communities, where fast food outlets and corner stores have proliferated, finding that locating in these neighborhoods remains profitable. This has created spatially unjust geographies of healthy food access within cities across the US (see Powell et al., 2007; Moore and Roux, 2006 and Morland et al., 2002).

Food vulnerabilities are concentrated among those with less wealth and political standing in our society. Instead of promoting self-reliance and community based food access options, the dominant food system creates and reinforces dependence on mostly large, profit driven entities that have little regard for the health of a given community and its residents. An urban political ecology perspective "contextualize[s] vulnerability at the local scale with any external or local pressure drivers that may have influence on access to nutritional and affordable food," and thus allows us to see how the structure and processes of the conventional food system produce food deserts and malnourished bodies at the local scale within marginalized urban communities (Stockholm Environmental Institute, 2004, p. 2). Another way of understanding vulnerability is as the absence of resilience within individuals and communities and that "resilience is often weakened by external or non placed based forces acting on the capacity of communities to cope with the prospect or actuality of problems" (Stockholm Environmental Institute, 2004, p. 2). Because most people are reliant/dependent on the external forces of global industrial agriculture and profit driven, placeless food retail businesses for healthy food access, their resilience to changes in food access, such as rising prices or the increasing absence of local food retail outlets, is weakened. Yet popular local food movement literature does not fully address the drivers of food vulnerability or the weakening of community resiliency. We identify two ways in which food vulnerability is produced: (1) through the structural mechanisms inherent within conventional food systems which work to exclude culturally sensitive produce and diets and (2) through a concurrent weakening in the capacity of low-income communities of color to access food (which is too expensive) and to consume healthy food in their neighborhoods. This reflects the dynamic between productions of vulnerability and the corresponding weakening of resilience. When analyzing the discourse surrounding local food systems, it is important to realize that the word "local" has multiple interpretations. As Born and Purcell (2006) have pointed out, scale is socially constructed and there is nothing inherent about any scale, including the local. Injustices are perpetrated and food vulnerabilities are produced at any scale; in other words actions at the local level in and of themselves do not guarantee a more sustainable and just food system. A small-scale local farmer for instance, is no more likely to use organic agriculture methods or pay a living wage to field workers than a large-scale export farmer half the world away. Furthermore, residents of disenfranchised neighborhoods are no more likely to have greater access to fresh and healthy produce solely due to the growth of a local food movement in an urban region if the movement is catering towards the more privileged segments of the city's population.

The framing of the local food movement in popular discourse has often confused the ends, which are a more sustainable and socially just food system, with the means; the localization of food production and consumption. In other words, the goal has become the creation of a local food system, rather than the creation of a more sustainable and just food system at the local level. Born and Purcell (2006, pp. 195–196) term this "the local trap," writing, "No matter what its scale, the outcomes of a food system are contextual: they depend on the actors and agendas that are empowered by the social relations in a given food system."

That being said, there are many examples of low income and underserved communities growing food locally, in urban areas that tend to have many vacant lots and underused spaces. Growing Power in Milwaukee, Wisconsin, the Detroit Black Community Food Security Network in Michigan, and Nuestras Raices in Holyoke, Massachusetts are successful examples of urban agriculture projects started within the communities that are fostering health for their neighborhoods. These local food systems create the potential for neighborhoods to supply themselves with more fresh and healthy produce, to address issues of food vulnerabilities, and build both economic empowerment/opportunity and community capacity for access, production and consumption. What is important to determine then, is who is being empowered by and benefitting from the localization of food, and who may be experiencing disempowerment or exclusion associated with these local food systems.

Politics and Heterogeneity within the "Local"

A second conflict in much of the local food system discourse is a denial of the diversity embedded in the "local." The assumption is that the localization of food will benefit all who take part, and those who are unable to participate or whom the localization process excludes largely go unrecognized. The "local" is constructed as discrete, homogenous, and static (Hinrichs, 2003). Diversity and unequal power relations exist within any given locality, and as noted above, certain actors and agendas may be empowered by and benefit from the uneven development of food system localization while others may be disempowered or left out.

In her research into the growth of the local food system in Iowa and the "Iowa-grown banquet meal," Hinrichs (2003) observed that localization can involve the defensive, exclusionary protection of a region with the assumption of homogeneity and common interests within that region and "otherness" outside. Yet, within any given region, unequal and contentious power relations can exist among the elite and the marginalized, the diverse ethnic and racial groups, and the different social classes. Localization may simply reinforce the agendas of the elite within a given region through the establishment of defensive protective territories for themselves at the expense of other local actors (Dupuis and Goodman, 2005). Localism can also serve the interests of family farms, restaurateurs, and their mostly white middle and upper income consumers while denying equal access to locally grown and culturally appropriate food for minority and low-income populations in a region. As Allen et al., (2003) demonstrated in their examination of alternative agrifood practice in California, leaders of the local food movement have expressed a clear preference for ecological sustainability over social justice in their discourses and agendas. Thus localism can deny the "politics of the local" and reflect the interests of a small, unrepresentative group, with potentially problematic social justice consequences (Dupuis and Goodman, 2005).

Allen and Hinrichs (2008) provide another example of exclusionary practice in their analysis of selective patronage "Buy Local" consumer campaigns. "Buy local" campaigns establish ambiguous boundaries for what is encompassed as the local, often delineated by state lines, such as is the case with the MassGrown "Buy Fresh, Buy Local" campaign in Massachusetts (MDAR). These market-based campaigns emphasize supporting local businesses and keeping consumer dollars circulating in the local economy, but often pay little or no attention to issues of social justice or food vulnerability for underserved communities within the local boundaries. There is little guarantee that the benefits of a buy local campaign will be distributed to the full range of producers and consumers within an ambiguous boundary, as only select farms will be listed in the campaign materials and directories and only some consumers will have the economic means to patronize them (Hinrichs and Allen, 2008).

Not only is there the danger of denying the politics within the local, there also exists the problematic exclusion of anything outside the local. The local – non-local binary can ignore the deleterious effects of localization on those who are placed outside the local boundaries. The classic example would be export agricultural farmers in developing countries who are denied access to markets in developed countries due to their local food preferences and exclusionary practices. In Cross et al.'s (2009) research into variances in farm worker health between localized and globalized food systems, they found that horticultural workers in Kenya's export oriented farms were often healthier than the population norm in their country, due to the stringent corporate responsibility practices of UK supermarkets that they supplied placed upon their production farms. The move towards more localized food systems in the UK will diminish the export market for Kenyan farmers, and some of these farms may go out of business, potentially resulting in a negative impact on the horticultural workers. The effects on those that are deemed outside the local or those that are not granted access to the local must be taken into consideration when attempting to build a more ecologically sustainable and socially just food system.

Finally, even those who do participate in the localization process are not necessarily better off because of it. As Jarosz (2008, p. 232) found in her examination of the local food system in western Washington state, "Increasing urban demand for seasonal and organic produce grown 'close to home' and the processes of rural restructuring which emphasize small scale sustainable family farms and its direct linkages to cities do not necessarily enable all farmers to consistently make a living from season to season." Jarosz found that many small-scale family farmers could not support themselves solely from selling produce directly to customers in urban markets and often had to take second jobs or have a spouse employed in another field. The inability to earn a living was primarily due to high transportation costs and competition among farmers within farmers' markets as well as from organic and health food groceries offering similar choices that drive down the prices urban consumers are willing to pay for their produce (Jarosz, 2008). Thus, those who are benefitting the most are the middle to upper income urban residents driving the demand for local produce who have a wide variety of choices in consumption, while some of the family farmers who are supposedly benefitting from the localization of food are having trouble making ends meet. New vulnerabilities are produced through the empowerment of certain actors within food system localization, such as middle to upper income urban consumers, while others are forced to adapt to less advantageous realities, such as the small scale farmer's supplying them. The local, however it may be defined, is never likely to be homogenous and static but rather heterogeneous and dynamic, with its own power relations, inequalities, and idiosyncrasies in which food system localization will have varying and selective beneficiaries as well as possible detrimental effects and exclusions that may further entrench existing food vulnerabilities. The key issue remains understanding who is empowered and who is disempowered by food system localization. The danger is that those who are most disenfranchised, made most vulnerable to the deleterious effects of the dominant food system, may be forgotten in the push for food system localization. As long as localization occurs, people may be convinced that positive social change is resulting, that all boats are being lifted, while in actuality, those who are left out are being ignored. Practices that allow for inclusive and democratic processes and results should be encouraged while those that are exclusive and perpetuate inequalities must be recognized and prevented.

Race, Culture, and Localization

Popular local food narratives have largely identified the ecologically and socially damaging forces of corporate power in the global food system as the primary object of struggle, while issues of race and racism have been given less attention (Slocum, 2006). The dominant discourse has been framed from a mostly white, middle and upper class perspective of food system activism. This locavore discourse is valorized in books such as Pollan's *Omnivores Dilemma* (2006) and the recent film *Food Inc.* (2009).

In Alkon's (2008) study of two East Bay farmers' markets, one located in a white, affluent neighborhood of Berkeley and the other located in a low-income, predominantly black neighborhood of West Oakland, she highlights the disconnect that can exist between different conceptions of local food: one (West Oakland) more influenced by a "just" sustainability agenda of equity and anti-racism, the other (Berkeley) more by an "environmental" sustainability agenda of reconnecting with nature. The socio-cultural divide between these two markets reflects a greater disparity amongst neighborhoods in the East Bay, and in Oakland in particular, between the lower-income, predominantly people of color neighborhoods of the "Flatlands," and the upper-income, whiter neighborhoods of "the Hills." The HOPE Collaborative, "a group of organizations, institutions, and community residents formed ... to improve health and quality of life by transforming the food and fitness environments in Oakland neighborhoods suffering the most from health disparities," conducted a study (2009) on supermarket access in the city.

The Collaborative found that in the Flatlands, where median household income is $32,219 per year and the population is 90 percent people of color, there is one supermarket per 93,126 people, while in the Hills, where median household income is $58,663 and the population in 48 percent people of color, there is one supermarket per 13,778 people (HOPE Collaborative, 2009). Due to the socio-economic and racial segregation between the Flatlands and the Hills, and the corresponding inequities in healthy food access, communities in the Flatlands are more susceptible to food vulnerabilities than the more affluent communities of the Hills. It follows then, that issues related to food vulnerabilities, such as racism and inequities, would be higher on the agenda at the West Oakland farmers market than at a market in a more affluent community, such as the one in Berkeley.

Alkon found through interviews, participant observation, and a review of the educational materials at both markets that the Berkeley market vendors and attendees were primarily concerned with the *environmental* sustainability agenda, placing emphasis on reconnecting with nature through local and organic farms, while the mostly African American West Oakland market vendors and attendees were primarily concerned with an *just* sustainability agenda that situated the local food movement as a response to racism and inequality (Alkon, 2008). Alkon (2008, p. 280) observed that, "Because the North Berkeley Farmers' Market works to improve the environment, defined as beautiful, diverse, non-human nature, some of its participants need not see the harsh realities of food insecurity. This inhibits the incorporation of social justice goals." The emphasis on either environmental sustainability or just sustainability reflects the racial and socio-economic complexion of the communities that the markets serve.

The two different sustainability paradigms "environmental" and "just," while generating different discourses of local food, are by no means exclusive of one another, but rather should "be seen as being both flexible and contingent, composed of overlapping discourses, which come from recognition of the validity of a variety of issues, problems and framings" (Agyeman, 2005). With respect to local food systems however, a "just" sustainability framework recognizes the power related issues of socio-economic and racial marginalization in linking sustainable food production with access to healthy, affordable, and culturally appropriate foods for racially and economically marginalized communities, as Alkon and Agyeman (2011) outline in *Cultivating Food Justice: Race, Class and Sustainability*. Many non-profit and community outreach organizations are now advocating for the establishment of community gardens, urban farms, farmers' markets, and food cooperatives in low income and minority communities to allow for greater availability of fresh and healthy produce to the community's residents and bridge the two agendas. However, advocacy for such proposals often arrives with certain cultural assumptions from proponents that may not always be entirely in line with those they are directed at.

As Slocum (2006) notes, the leadership of community food security organizations tend to be heavily white, while those they are attempting to serve, who disproportionately experience food vulnerabilities and are often from communities of color, tend to be the subjects, rather than the leaders of such organizations. The directors and managers of these organizations often come from well-educated backgrounds with advanced degrees, and thus from positions of privilege, yet there is a lack of recognition of this privilege and more specifically of white privilege. These organizations are operating within the bounds of institutionalized racism in which those who are advocating for community food security are often able to do so only because of their positions of privilege, which allowed them to access the education and status that has set them apart from those they are trying to serve. Slocum (2006) calls for a reframing of the movement with an "anti-racist" lens that recognizes institutionalized racism within the local food and community food security movements and seeks to address the inequalities and privileges that exist not only in the food system but within the movement itself.

Due to the dominant framing of the discourse from a white, middle class perspective, much of the work to address food vulnerabilities being conducted reflects white cultures of food and white histories that may be culturally insensitive to those being served (Guthman, 2008). Guthman has written extensively on this subject through her observations of food vulnerabilities work in California. She has observed how many of the food vulnerabilities initiatives targeting underserved urban minority communities are met with less enthusiastic responses and participation rates than the organizers anticipate (Guthman, 2008). These initiatives often involve the establishment of urban farms or community gardens and encourage community residents to grow their own produce and provide food security for themselves. Yet, sometimes, those being targeted have specific cultural histories that see farming and growing food as reminders of past oppression. This especially can be the case with African American communities whose ancestors have had a tormented past of slavery and share cropping in the US and whose residents may want little to do with growing their own food. An act seen as positive and empowering from a local food perspective reflecting white cultural histories can be perceived as an unwanted reminder of past injustices from a black cultural perspective. As Guthman (2008) noted, some of the residents in the African American communities she observed would rather have had a Safeway locate in their neighborhood than an urban farm. These residents simply want the same privilege that most middle class Americans enjoy, being convenient access to grocery stores, and were uninterested in more alternative forms of food production and consumption.

There are, however, underserved minority populations, including refugees, that have welcomed the opportunity to participate in urban agriculture and local food projects that allow them to grow culturally appropriate food, which may be unaffordable or unavailable in grocery stores and farmers' markets otherwise (Agyeman 2011). Some examples are African immigrants in the Washington DC/ Maryland area, Latino communities in cities such as Los Angeles and New York, as well as other immigrant populations with strong agrarian histories, who have been taking advantage of local food and urban agriculture initiatives to continue their agrarian and culinary traditions in the US (Saldivar-Tanaka, 2004). One notable example is the New Entry Sustainable Farming Program administered by the Friedman School of Nutrition at Tufts University, which provides recent immigrants with assistance in starting small scale farming operations in Massachusetts (New Entry website).

As we have shown, the mantra of the popular locavore discourse does not resonate equally among all communities. Implicit in the popular discourse of much of the local food agenda is the dominance of white cultural histories, narratives and white privilege. For the local food movement to become truly transformative in addressing food vulnerabilities issues, power dynamics, institutionalized racism, classism and cultural insensitivity must be recognized and confronted, as they are part of the production of food vulnerabilities (Heynen and Swyngedouw, 2006). As long as race is overshadowed by an emphasis on "environmental" sustainability concerns, as in the Berkeley case, or there is a focus on addressing food vulnerabilities solely in terms of social class, then the potential exists for certain racial and cultural groups to be marginalized or excluded. Furthermore, not only do popular notions of local food activism create the potential for exclusion, they also reinforce the hegemonic economic and political paradigms that have fostered the dominant food system which the local food movement is supposedly attempting to counter and thus reinforce patterns of food vulnerability ... (see Allen and Guthman, 2006; Dupuis and Goodman, 2005 and Guthman, 2008).

Building Resiliency within the Local

If local food movements are to truly foster a more socially just food system, then they must recognize and address the production of food vulnerabilities. As stated earlier, vulnerability is produced and maintained in communities that lack resilience. We have argued that communities lacking self determination in producing and consuming food are made vulnerable to the external processes and power dynamics of the global capitalist industrial agro-food system, in which fluctuations in commodity prices and capital flows can determine a community's access to healthy food, rates of malnourishment increased risks for cardiovascular disease. To address the production of food vulnerabilities, the local food movement, or any alternative food movement for that matter, must begin to build food resiliency within communities. Folke et al., (2003, p. 355) described the four factors that are essential to resiliency in socio-ecological systems:

1. learning to live with change and uncertainty;
2. nurturing diversity for reorganization and renewal;
3. combining different kinds of knowledge; and
4. creating opportunity for self-organization.

Local food systems may have the potential to allow food vulnerable communities to meet these criteria. Buchmann (2009) has documented how home gardeners in Cuba have helped to strengthen community resilience through meeting Folke et al.'s four criteria. By growing food for themselves and their neighbors they are learning to accept the volatility of food access in the isolated Cuban food system, nurturing a diversity of plants grown for both consumption and medicinal use within and amongst home gardens, using traditional knowledge of the local ecology as well as modern scientific knowledge, collectively sharing resources, knowledge, and skills across their communities, and allowing for flexibility and self-organization in the formation of supportive social networks (Buchmann, 2009). Buchmann concludes that, "economic crises and frequently changing policies on agriculture, food security, religious freedom and health care have had an impact on household decision making, and therefore ... on home garden composition and management," and that, "local social networks developing in and around home gardens form the backbone of local community resilience" (2009, p. 719).

Although the context of food vulnerabilities is entirely different for underserved urban communities within the US, the lessons learned from Cuban home gardeners apply. Local food systems can begin to counteract the production of food vulnerabilities and build resilient communities through developing a greater diversity of food sources beyond traditional food retail outlets, such as home and community gardens, urban agriculture, community food cooperatives, and farm shares; encouraging the production and consumption of a greater diversity of food crops; combining local community and cultural knowledge with local agricultural production and ecological knowledge; and by creating self-organized, mutually supportive social networks outside of traditional market relations.

Re-imagining the Local

The localization of food production should be recognized as one means toward a more socially and environmentally sustainable food system, not an end. As we have argued producing and consuming locally does not guarantee any greater concern for social justice or more ecologically sustainable practices, unless those principles are internalized into the local food movement, including food policy councils (Hamilton, 2002; Harper et al., 2009) and local not-for-profits. Localization does however hold the potential to make democratic decision-making more transparent, understandable and achievable by involving a greater number and diversity of stakeholders who can experience the effects of their decisions in a more direct manner. Decentralizing production and rooting food sources in the communities that they serve can reduce dependence on the exogenous processes of the capitalist industrial food system. Localization can also provide for greater social connectedness in food system processes due to the physical and relational proximity between producers and consumers. Furthermore, as Anderson (2008, p. 602) notes, "Localization allows greater food system diversity because each locale can support unique foodways and a unique set of relationships between producers and buyers."

Yet none of these enabling processes of local food systems guarantee a more socially just result. First, there must be recognition of the multi-scalar socio-cultural, economic, and ecological dynamics that manifest themselves in highly uneven power asymmetries and built environments thereby producing food vulnerabilities, especially for low income and minority populations. Second, the fostering of *food resiliency* to reverse the production of food vulnerabilities must be an intentional and explicit goal of local food systems if a more socially just result is to be possible.

In striving for social justice in food system advocacy, we must also recognize that food is heavily culturally located. Different ethnicities and cultures experience food in myriad, non-conforming ways. Each group has their own unique cultural understandings and practices of food. Any attempt to build social justice through food system activism must take a multi-lens approach that recognizes the diversity of cultural experiences and histories involved. Food system activists must acknowledge the hegemonic power of the white cultural framings of the popular food system activism and the existence of white privilege within the local food movement, and attempt to reconcile these inequalities through cultural sensitivity and anti-racist framings. Food system activism and food policy councils must become more socially inclusive and democratic, involving the greatest participation of stakeholders possible. By taking a multi-lens approach, the heterogeneity and unequal power dynamics of the local can be accounted for.

Finally, recognition of human rights, or a shared responsibility to ensure good food opportunities for all, in both production and consumption, must be the overriding objective if we are to develop a more socially just, environmentally sustainable and resilient food system. The concept of food vulnerabilities offers a promising framework for re-imagining the local, as it highlights the multiplicity of processes, forms of and sites of food vulnerability and the ability of underserved, marginalized, and diverse communities to produce, access, and consume healthy and culturally appropriate food.

Whatever scale of solution food system activists advocate for, whether it be a local food production system informed by the concept of food vulnerabilities or the building of international solidarity movements toward global food sovereignty (Desmarais, 2008), understanding the socio-economic, cultural, and ecological contexts in which solutions are being applied is of the utmost importance. Fostering resilient communities, those that are empowered to produce and access healthy, fresh, and culturally appropriate food in a sustainable manner, must be the ultimate goal. Only by recognizing the diversity of histories, cultures, narratives and objectives at play, and through establishing a dialogue amongst the range of interests to negotiate a common ground, can a truly inclusive and representative solution be imagined.

References

Agyeman, J. (2011). "New agricultures, cultural diversity and foodways," <http://julianagyeman.com/2011/10/new-agricultures-cultural-diversity-and-foodways/>. Accessed November 2 2011.

Alethea, H., Shattuck, A., Holt-Giménez, E., Alkon, A. and Lambrick, F. (2009). Food policy councils: lessons learned. Food first: Institute for Food and Development Policy.

Alkon, A. (2008). "Paradise or pavement: the social constructions of the environment in two urban farmers' markets and their implications for environmental justice and sustainability," *Local Environment*, vol. 13, no. 3, pp. 271–289.

Alkon, A. and Agyeman, J. (eds) (2011). *Cultivating Food Justice: Race, Class and Sustainability*. Cambridge: MIT Press.

Allen, P. (2008). "Mining for justice in the food system: perceptions, practices, and possibilities," *Agriculture and Human Values*, vol. 25, no. 2, pp. 157–161.

Allen, P. (2010). "Realizing justice in local food systems," *Cambridge Journal of Regions, Economy and Society*, vol. 3, pp. 295–308.

Allen, P., FitzSimmons, M., Goodman, M. and Warner, K. (2003). "Shifting plates in the agrifood landscape: the tectonics of alternative agrifood initiatives in California," *Journal of Rural Studies*, vol. 19, no. 1, pp. 61–75.

Allen, P. and Guthman, J. (2006). "From 'old school' to 'farm-to-school': neoliberalization from the ground up," *Agriculture and Human Values*, vol. 23, no. 4, pp. 401–415.

Allen, P., Guthman, J. and Morris, A.W. (2006). "Squaring farm security and food security in two types of alternative food institutions," *Rural Sociology*, vol. 71, no. 4, pp. 662–684.

Allen, P. and Hinrichs, C.C. (2008). "Selective patronage and social justice: local food consumer campaigns in historical context," *Journal of Agricultural and Environmental Ethics*, vol. 21, no. 4, pp. 329–352.

Anderson, M.D. (2008). "Rights-based food systems and the goals of food systems reform," *Agriculture and Human Values*, vol. 25, no. 4, pp. 593–608.

Baker, L.E. (2004). "Tending cultural landscapes and food citizenship in Toronto's community gardens," *Geographical Review*, vol. 94, no. 3, pp. 305–325.

Born, B. and Purcell, M. (2006). "Avoiding the local trap: scale and food systems in planning research," *Journal of Planning Education and Research*, vol. 26, no. 2, pp. 195–207.

Buchmann, C. (2009). "Cuban home gardens and their role in socio-ecological resilience," *Human Ecology*, vol. 37, no. 6, pp. 705–721.

Community Food Security Coalition, <http://www.foodsecurity.org>.

Cross, P., Edwards, R.T., Opondo, M., Nyeko, P. and Edwards-Jones, G. (2009). "Does farm worker health vary between localised and globalised food supply systems?" *Environment International*, vol. 35, no. 7, pp. 1004–1014.

Desmarais, A.A. (2008). "The power of peasants: reflections on the meanings of La Vía Campesina," *Journal of Rural Studies*, vol. 24, no. 2, pp. 138–149.

Detroit Black Community Food Security Network, <http://detroitblackfoodsecurity. org/>.

Diex Roux, A., Morland, K. and Wing, S. (2006). "Supermarkets, other food stores, and obesity: the atherosclerosis risk in communities study," *American Journal of Preventive Medicine*, vol. 30, no. 4, pp. 333–339.

DuPuis, E.M. and Goodman, D. (2005). "Should we go 'home' to eat?: toward a reflexive politics of localism," *Journal of Rural Studies*, vol. 21, no. 3, pp. 359–371.

Folke, C., Colding, J. and Berkes, F. (2003). "Synthesis: building resilience and adaptive capacity in social-ecological systems." In: F. Berkes, J. Colding and C. Folke (eds) *Navigating Social–Ecological Systems: Building Resilience for Complexity and Change*. Cambridge: Cambridge University Press, pp. 352–387.

Food Inc. (2008). [Film] Directed by Robert Kenner. USA: Participant Media.

Giang, T., Karpyn, A., Laurison, H., Hillier, A., Burton, M. and Perry, D. (2008). "Closing the grocery gap in underserved communities: the creation of the Pennsylvania fresh food financing initiative," *Journal of Public Health Management and Practice*, vol. 14, no. 3, pp. 272–279.

Growing Power, Inc., <http://www.growingpower.org/>.

Guthman, J. (2008). "Bringing good food to others: investigating the subjects of alternative food practice," *Cultural Geographies*, 15, pp. 431–447.

Guthman, J. (2008). "Neoliberalism and the making of food politics in California," *Geoforum*, vol. 39, no. 3, pp. 1171–1183.

Hamilton, N.D. (2002). "Putting a face on our food: how state and local food policies can promote the new agriculture," *Drake Journal of Agricultural Law*, vol. 7 no. 2, pp. 408–454.

Heynen, N. (2006). "Justice of eating in the city: the political ecology of hunger," In: N. Heynen, M. Kaika and E. Swyngedouw (eds) *In the Nature of Cities: Urban Political Ecology and the Politics of Urban Metabolism*. London: Routledge, pp. 129–142.

Heynen, N. and Swyngedouw, E. (2003). "Urban political ecology, justice and the politics of scale," *Antipode*, vol. 35, no. 5, pp. 898–918.

Hincrichs, C.C. (2003). "The practice and politics of food system localization," *Journal of Rural Studies*, vol. 19, no. 1, pp. 33–46.

HOPE Collaborative (2009). "A place with no sidewalks: an assessment of food access, the built environment and local, sustainable economic development in ecological microzones in the city of Oakland, California in 2008," (preliminary findings) cited in Treuhaft, S., Hamm, M.J. and Litjents, C. (2009). "Healthy food for all: building equitable and sustainable food systems in Detroit and Oakland," PolicyLink and Michigan State University.

Jarosz, L. (2008). "The city in the country: growing alternative food networks in metropolitan areas," *Journal of Rural Studies*, vol. 24, no. 3, pp. 231–244.

Kingsolver, B. (2007). *Animal, Vegetable, Miracle: A Year of Food Life*, New York: Harper Perennial.

Moore, L. and Roux, A. (2006). "Associations of neighborhood characteristics with the location and type of food stores," *American Journal of Public Health*, vol. 96, pp 325–331.

Morland, K., Roux, A. and Wing, S. (2002). "The contextual effect of the local food environment residents'diets: the atherosclerosis risk in communities study," *American Journal of Public Health*, vol. 92, no. 11, pp. 1761–1767.

New Entry Sustainable Farming Project, <http://nesfp.nutrition.tufts.edu/index.html>.

Nuestras Raices, <http://www.nuestras-raices.org/>.

Pollan, M. (2006). *Omnivore's Dilemma: A Natural History of Four Meals*, New York: Penguin.

Powell, L., Slater, S., Mirtcheva, D., Bao, Y. and Chaloupka, F. (2007). "Food store availability and neighborhood characteristics in the United States," *American Journal of Preventive Medicine*, vol. 44, pp. 189–195.

Saldivar-Tanaka, L. and Krasny, M.E. (2004). "Culturing community development, neighborhood open space, and civic agriculture: the case of Latino community gardens in New York City," *Agriculture and Human Values*, vol. 21, no. 4, pp. 399–412.

Slocum, R. (2006). "Anti-racist practice and the work of community food organizations," *Antipode*, vol. 38, no. 2, pp. 327–349.

Stockholm Environmental Institute (2004). "Political ecology of vulnerability," SEI Oxford/GECAFS Methodological Briefs, pp. 1–4, available at: <http://www.gecafs.org/publications/index.html>.

The Food Project, <http://thefoodproject.org/>.

Wrigley, N. (2002). "Food deserts' in British cities: policy context and research priorities," *Urban Studies*, vol. 39, no. 11, pp. 2029–2040.

Chapter 6

Sustainability Planning, Ecological Gentrification and the Production of Urban Vulnerabilities

Sarah Dooling

Introduction

In the current political landscape of the United States, homeless people and immigrants – in particular Mexican immigrants – represent two vulnerable social groups that are lightening rods for politicians and communities involved in planning the future of cities. During this economic recession, with a national unemployment rate hovering between 9 and 10 percent (Brookings Institute, 2010), homeless individuals are being targeted as a drain on limited fiscal resources, and budgets for services aiding homeless people are being significantly cut with federal and state shortfalls (Liberto, 2011; Fears, 2009). The psychological, political, and economic consequences of being homeless have been extensively addressed in the literature (Scott, 2007; Arnold, 2004; Desjarlias, 1997; Feldman, 2004; Hopper, 2003; May, 2000; Rowe and Wolch, 1990; Ropper, 1988). As people who lack access to safe, predictable housing, homeless individuals are forced to enact their private daily lives in public, thus blurring the boundaries between public and private (Dooling, 2009; Waldron, 1991). They remain constantly at risk of violating civility ordinances that make it illegal to sit or sleep on sidewalks, or camp in parks – behaviors that constitute daily actions of homeless people (Gibson, 2004). The symbolically powerful images of dirtiness, criminality, violence, drug addiction and laziness are considered to reflect and constitute a homeless person's individual moral failing. The emphasis on individual agency does not acknowledge how the larger scaled structural changes in economic production, lack of universal health care, insufficient employment wages, and inadequate supply of inexpensive housing options are major drivers of urban homelessness (Burt et al., 2004; Hopper, 2003; Koegel et al., 1996; Rossi, 1989). Homelessness and homeless people in particular, are framed from deeply politicized positions driven, in part, by (1) the idea that homeless people are morally deficient and are therefore responsible for their plight and (2) the uncanny experience of housed residents encountering homeless people creating *private home-like spaces illegally* in *public*. Stigmatization of homeless individuals and homelessness as an unwanted urban phenomenon, contributes to polarized positions, while ineffective politics continue to create conditions of vulnerability under which homeless people persist.

The movement of immigrants, especially undocumented workers, into the US (and especially in the South West) has been greeted with similarly entrenched political positions. Conservatives assert that as illegal aliens in the US, these individuals – many of the families – are draining limited resources that should more rightfully be allocated to legal US citizens (Schrag, 2010). The failure to pass the DREAM Act[1] and the passage of the toughest state immigration law in the US by Arizona voters[2] reflect such political ideology – i.e., illegal status should limit, if not deny, a person's legitimate political and economic participation in US society, and illegal residents risk incarceration and deportation. Others assert that undocumented workers provide a critical source of labor, accepting wages and jobs considered unacceptable to other social groups; they argue that excluding this labor pool from the US economy – particularly in California, Arizona and Texas – could have disastrous effects (Farrell, 2010). Associated with this debate is the increasing awareness of the violence along the US–Mexico border related to the drug trades and drug cartels (Bowden, 2010; Campbell, 2009). Perceptions of Mexican immigrants as violent, drug addicted, uneducated criminals contribute to the polarized character of the public discourse surrounding immigration in the US (Streitmatter, 1999).

Both undocumented immigrants and homeless people are enacting lives within the cracks between the binaries of public and private, legal and illegal. The blurring of boundaries – public/private, legal/illegal – contributes to the polarizing effect that perceptions of each group has on political debate, on-going public dialogue, and planning efforts aimed at creating cities that are economically productive, socially vibrant and environmentally healthy. For some, homeless people and undocumented Mexican immigrants are viewed as contributing to crime in US cities and acting as financial drains on the local and state economies despite evidence to the contrary (Turnbull, 2009). There is also media coverage that portrays the environmental impacts of homeless people sleeping outside in parks and green spaces in terms of public health threats from the accumulation of drug paraphernalia, trash and lack of personal hygiene practices (Bawarshi et al., n.d.).

---

1   The DREAM Act was sponsored by Senator Richard Durbin and Representative Howard Berman in 2009. The purpose of the Development, Relief and Education of Alien Minors Act was to help individuals who meet certain requirements be able to enlist in the military or go to college and obtain citizenship which they otherwise would not have without this legislation. It specifically targeted undocumented immigrant students who have been living in the US since they were young. The proposed legislation was defeated by the Senate in December 2010.

2   In April 2010, the Arizona Governor signed the most stringent law related to the identification, prosecution and deportation of illegal immigrants. This law makes the failure to carry immigration documents a crime and gives the police broad powers to detain individuals suspected of being illegal.

Immigrant populations are portrayed as failing to take care of their neighborhood through the proliferation of trash and broken down vehicles (Streitmatter, 1999). Thus the polarized (and polarizing) discourse can be organized as: (1) immigrants and homeless people create economic, cultural and environmental harms, and (2) simultaneously, they themselves are vulnerable to persistent economic hardship, arrest or deportation, and to being portrayed as a risk in and of themselves to society at large.

In this chapter I analyze the impacts of sustainability planning on homeless people camping in land designated as an urban ecological reserve and on Mexican immigrant households living in a neighborhood targeted for a transit-oriented, mix-use development. This comparative analysis serves multiple purposes:

1. to compare the unequal distribution of benefits to the goals and components associated with the respective sustainability plans;
2. to identify the actual and potential harms generated by sustainability plans for already vulnerable groups of people; and
3. to demonstrate the contradictions associated with sustainability planning which espouses maximization of ecological, economic and equity issues in the transformation of cities.

Sustainability planning is one strategy among the design allied professions intended to address the vulnerability of urban areas to environmental and economic changes (Portney, 2003; Newman and Kenworthy, 2003), with particular focus paid to the threats posed by climate change. Policy makers and planning agencies recognize increased concentrations of greenhouse gas emissions being a function of segregated land uses, transportation infrastructure and patterns of travel behavior (Ewing et al., 2008). Planners are interested in enhancing the overall urban ecological capacity of cities, including the capacity to sequester carbon, connect wildlife habitat and increase the amount of impervious cover to alleviate flooding and water contamination problems (Birch and Wachter, 2008; Forman, 2008). Sustainability planning – including the designation of urban ecological reserves and transit-oriented developments – is intended to minimize and mitigate urban ecological vulnerabilities.

This chapter analyzes how urban vulnerabilities are produced through the collusions of environmental and economic agendas that fail to address existing conditions of vulnerable people who are stigmatized and vilified in the popular media. Planning agendas work alongside media portrayals to exacerbate existing and set up new conditions of vulnerability for – in the two cases presented here – homeless individuals camping in greenbelts and low-income immigrant households living in a neighborhood undergoing a transit-oriented redevelopment effort. In addition, the transit-oriented planning effort for a predominantly immigrant neighborhood reveals the ways in which social vulnerabilities contribute to the undermining of ecological functionality, and perpetuating the city's future risks to the impacts associated with climate change.

Ecological gentrification is a useful concept out of which convergences between ecological and economic agendas, and the resulting contradictions, can be identified and described. Broadly defined, ecological gentrification is the uneven distribution of benefits associated with a planning effort driven by ecological agendas or environmental ethics (for a wider discussion on ecological gentrification see Dooling, 2009). The term is intentionally provocative, as it seeks to associate displacement of people (as a process typically associated with economic development and neighborhood change) with environmental changes stemming from formalized planning efforts, where the changes are assumed to be and referred to as universally beneficial. Ecological gentrification offers a critical perspective on the dynamics between social and environmental change that is based upon dialectical analyses of multi-scalar (from neighborhood to regional) interactions in cities. Referred to elsewhere as eco-gentrification (Quastel, 2009), the concept challenges core assumptions about the universal value of enhancing ecosystem function when placed in larger social contexts. The contradictions include creating new and exacerbating existing risks and harms for groups of people that are least able to avoid these risks and mitigate associated harms. The contradictions are result of the inevitability of co-occurring social and environmental changes and fragmented planning approaches that do not pay attention to these dialectical dynamics. I present two ways in which ecological gentrification operates in the urban context to produce urban vulnerabilities, and discuss the conceptual linkages between ecological gentrification and the production of urban vulnerabilities in relation to sustainability planning efforts.

Urban Homelessness and Green Space Planning in Seattle

Seattle, Washington is known as the Emerald City (Gibson, 2004), owing in part to the legacy of the Olmsted Parks Plan of 1903 that contributed to the current pattern of approximately 10 percent of city owned land designated as city-owned parkland (Dooling et al., 2006). One of the earlier established park preserves (Schmitz Preserve), designated in 1912, reflected an early environmental consciousness that understood the impacts of excessive timber extraction on people's perceptions about the city's beauty and ecological health (Dooling et al., 2006). As the first US city to be home to a federally listed endangered salmonid species in 1993 (Klingle 2010), Seattle municipal agencies are mandated to improve riparian habitat quality for migrating salmonid species; state growth management regulations, passed in 1990 and 1991, attempt to minimize the impact of urban growth development on ecosystems through establishing an urban growth boundary that targets future growth in already densely populated areas with existing infrastructure (Dooling et al., 2006; Robinson, 2005).[3] Greenspaces, including parks, are perceived as

---

3   Research has demonstrated that the effectiveness of the urban growth boundary has been questionable, and sprawling low density development outside the UGB occurred, thus increasing habitat fragmentation (Robinson et al., 2005).

valuable urban amenities, serving both social purposes (i.e., active and passive recreation) and ecological purposes (i.e., habitat, carbon sequestration). In 1993, Seattle adopted a Green Space Policy that led to the identification of 30 city properties targeted for preservation in order to protect natural habitat, mitigate noise and air pollution, reduce stress on aging water infrastructure and preserve sources of natural drainage. Greenspace, as an officially designated type of parks department property, is defined as land set aside to protect natural amenities and ecological functioning (Seattle Parks and Recreation Department 2000).

Seattle's environmental consciousness, progressive state regulations and city ordinances aimed at protecting wildlife species and enhancing their habitats appear to be starkly different, and distant, from some of the critical social issues the city currently faces. Seattle is a city known for its lack of affordable housing (Gibson 2004). Although home prices have declined over the past four years (the 2006 median house price was half a million dollars {Shrestha, 2007} and 2010 median price for a single family home was $399,000 {Cohen, 2011}), housing affordability remains a concern. The 2008 housing wage for Seattle/King County was 224 percent of minimum wage, more than double the 2008 minimum wage of $8.07 (National Low Income Housing Coalition, n.d.). The historical demolitions of single-room occupancy hotels (Roper, 1988; Groth, 1994; Gibson, 2004), which once provided inexpensive housing for seasonal and low-wage earners, and the rapid rates of converting apartments to condominiums (Cohen, 2008), also contributes to the lack of affordable housing. In conjunction with poverty, the lack of affordable housing is considered to be a leading contributor to the production of homelessness (Shinn and Gillespie, 1994; Koegel et al., 1996).

For a city with 563,374 inhabitants, Seattle's homeless population constitutes less than 1 percent. In a 2009 survey based on interviews with 89 homeless individuals, 70 percent had been homeless for more than one year and only 37 percent reported using a shelter within the past six months (United Way et al., 2009). Between 1999 and 2004, the homeless population was estimated to have increased 40 percent, peaking at an estimated 8,300 homeless people within King County (Seattle/King County Coalition for the Homeless, 2005). On average, American cities have reported a 12 percent increase in homeless populations since 2007 (US Conference of Mayors, 2008). These increases come in the wake up local planning efforts (known as the Ten Year Plans to End Homelessness) around the country which were having significant impacts: between 2005 and 2007, the National Alliance to End Homelessness estimated a 10 percent reduction nationwide (National Alliance to End Homelessness, 2009). In Seattle, there was a 5 percent decrease in the numbers of homeless individuals counted in 2010 (2,675) compared to 2009 (2,827) for King County; numbers of homeless individuals counted in Seattle was the same between 2009 and 2010 (1,986) (Seattle/King County Coalition on Homelessness, 2010). Some of this decrease is partly attributed to the federal initiative to end homelessness. Currently, Seattle, like over 290 communities in the US, is in the process of developing and implementing what is referred to as *The Ten Year Plan to End Homelessness*.[4]

However, despite the small percentage of homeless individuals in Seattle's population, these people are highly visible because they are forced to enact their daily private lives in public, including sidewalks, alleys and city parks (Dooling, 2009; Waldron, 1991). Homeless people, in living their private lives in public spaces, destabilize and confuse the boundaries between public and private, and between park and home (Dooling, 2009). In addition, many of them have (on average) a reduced life expectancy (O'Connell, 2005) due to increased exposure to elements, illness (e.g., occasional outbreaks of tuberculosis in homeless shelters) and street violence (Anderson, 1996). These data indicate homeless people live in tenuous, unstable and unsafe conditions, and that local and national forces (i.e., economic recession, inadequate wages, and insufficient amount of and access to affordable housing) work in concert to maintain their vulnerable condition. However, this conventional political economic assessment of homelessness and homeless people has long been documented. There is now a new twist emerging, and its origins reside in the environmental community and in sustainability planning efforts (Dooling 2008, 2009).

Green space, as an official park land designation, is defined as land set aside to protect natural amenities and ecological functioning (Seattle Parks and Recreation Department, 2000). The West Duwamish Greenbelt, as one example of an officially designated green space, is over 500 acres and represents the largest remaining contiguous forest within Seattle city limits (Seattle Parks and Recreation Department, n.d.). Site maintenance, much of it conducted by local non-profits, consists of removal of non-native vegetation and re-planting of native conifers. Kiwanis Ravine Overlook, another designated green space, is close to a well-used and large park in a Seattle neighborhood, north of the Duwamish Greenbelt. As an urban forest, the Ravine supports an active rookery of Great Blue Herons. An active forest restoration effort, including the removal of non-native plant species and the planting of native plants and trees, is on-going.

Yet, these places, as well as many other greenbelts throughout the city, are also where homeless individuals camp. Previous research based on interviews with

---

4   The US Department of Housing and Urban Development (HUD) has mandated that ten year plans be developed by communities in order to receive HUD money for homeless services. HUD rationalizes the local approach to policy development by claiming homelessness varies locally and, consequently, effective strategies to "end homelessness" need to be developed locally. The role of HUD in the development of ten year plans is primarily facilitative, providing technical expertise to local policy development groups. Critics of HUD's *Ten Year Plan to End Homelessness* initiative assert that HUD is devolving its responsibility for creating a national housing strategy and, with concurrent budget cuts in housing voucher programs and homeless services funds, critics also assert that HUD is setting up local policy development efforts to fail through lack of adequate funding (Western Regional Advocacy Project, 2006).

Seattle homeless people revealed that many individuals, in particular men, seek out wooded parks and greenbelts not because they prefer sleeping outside, but because all other options provided them are untenable: shelters are experienced as spaces of crime, violence, disease and religious indoctrination; and single room occupancy residences restrict the frequency and number of guests and, in some cases, restrict the use of appliances (Dooling, 2009).[5] For some, including Mel, a Native American, living in a single-room occupancy unit did not allow him to socialize with his close group of friends, and the disruption of his social network proved too difficult for him to retain housing. For others, the lasting effects of profound childhood abuse made living in a house – a structure which conjured up traumatic experiences – made it impossible to be contained within conventional housing. Still others, because of work-related injuries and the eventual loss of employment-subsidized health care, sleep in green spaces because it affords them privacy and refuge compared to shelters and transitional housing options. Green spaces are thus places where homeless people camp in order to maintain a sense of autonomy and privacy, to recover from traumatic experiences, and to maintain social connections that are not supported by conventional housing regulations.

In 1993, Seattle City Council approved a series of ordinances known as the Civility Ordinances (Gibson, 2004). The no-sitting ordinance prohibits sitting on any sidewalk between 7:00 a.m. and 9:00 p.m. and carries a $50 fine and possible incarceration (Bass, 2000). In 1997, City Council approved a parks exclusion ordinance in response to complaints that parks were becoming homeless campsites. Individuals caught drinking, camping or committing "acts of misbehavior" are banned from parks. Individuals with repeat violations can be banned from parks by police for up to one year (Bass, 2000). Many scholars assert that ordinances regulating behavior in public – including no-sitting and anti-camping ordinances – target and criminalize homeless individuals in efforts to create clean and safe spaces for housed urban residents (Marin, 1987; Waldron, 1991; Zukin, 1995; Mitchell, 2003; Arnold, 2004; Feldman, 2004). One can expand this argument by claiming that ideological notions about the primacy of urban ecological spaces also contribute to conditions of vulnerability – in particular to the risk of arrest, expulsion and banishment – experienced by homeless people sleeping in green spaces. The parks department definition of green spaces conceptually excludes homeless campers by virtue of the land use designation – whereas housed residents are encouraged to participate in the maintenance of these greenbelts the parks department actively expels homeless campers through the enforcement of the anti-camping ordinance.

While ordinances that establish the illegality of sleeping in parks are not new, their enforcement in the service of an ecological agenda is. Establishing and maintaining public green spaces at the exclusion of planning for alternatives for homeless campers demonstrates how valuing of public green spaces for ecological

---

5   One individual interviewed spoke about the oven in his SRO unit restricted to being used for only 30 minutes.

purposes contributes to intensifying vulnerable conditions of an already vulnerable group of people. Otherwise commendable civic goals of improving habitat quality and connectivity for urban wildlife species, increasing pervious land cover for protecting groundwater and mitigating flooding, and enhancing the city's capacity for carbon sequestration dominate the public discourse at the expense of citizens who are most directly affected by exclusionary aspects of the land preservation policy. More bluntly, the implementation of an ecological agenda, coupled with regulatory strategies to enforce the ecological values of the landscape, produces harm for the poorest people. Ecology, supported and enforced through regulatory practices, emerges as a moral authority with material and legal consequences for homeless people.

Neighborhood Transit Oriented Development Planning in Austin

Austin, Texas is another city known for its environmental awareness and commitment to issues of sustainability (Swearingen, 2010). Recently, the City of Austin hired a Sustainability Officer (from Seattle) to work towards developing a city-wide vision related to energy consumption, carbon sequestration, water conservation, and integrated land use patterns (Gregor, 2010). During the 1990s, the city focused its efforts on shifting development from the west side of town to the east, to limit development over the ecologically sensitive karst aquifer that serves as a source of the city's drinking water. In addition, development downtown has been promoted under Smart Growth and sustainable development rationales. At the same time, a series of initiatives have focused on fostering development along "core transit corridors" and mixed use, higher intensity development near transit stations.

The historical context for this redevelopment is the pattern of racial and economic segregation set in motion through the city's 1928 comprehensive plan. In this plan, all public facilities for African-Americans were restricted to a neighborhood just east of the central business district, resulting in a pattern of segregation that persists. The Latino community, while not formally limited to a similar zone, gradually formed a concentrated community just south of the "negro district." The East Riverside corridor runs along the edge of one historically low income Latino neighborhoods and its large number of subsidized housing units is indicative of its socio-demographic position in the city.

The East Riverside Master Corridor (ERMC) plan emerged from previous discussions about strategies for development of a light rail network for the city. The main arterial in the neighborhood, Riverside Drive, was the recommended route: it connected key destinations, was predicted to maximize ridership, and offered strong potential for value capture through transit-oriented development and offering a wide roadway (City of Austin, 2009d). In addition, planning staff pointed to two other reasons for transforming the neighborhood, including the high concentration of rental housing in the zone closest to the highway (the area most attractive for redevelopment) and crime, which was often associated with the area's rental housing. Local homeowners favored redevelopment that would increase the share of homeowners in the area (Ng, Grant, Donoso and Eaton, 2007). The plan was framed around the physical transformation of the neighborhood aligned with city-wide transit goals and tied to redevelopment agendas intended to economically and ecologically revitalize the neighborhood. Little was known about the views of the numerous low-income immigrant renters living in the community, and there was no charge from city council to the consultant to assess the population's needs nor how these needs relate to the goals presented for the plan.[6]

The broader public entered the process when a "visioning workshop" was held in September 2008, which was attended by nearly 150 people.[7] Participants completed a Visual Preference Survey and Community Questionnaire that was intended to evaluate how participants valued various elements of the current community and assessed alternatives.[8] (An additional 450 completed the questionnaire online.) Not surprisingly, given the cost-free choice of selecting the most visually appealing images, the end result was an endorsement of the initial charge for the plan: "an area that welcomes and encourages pedestrian activity by providing a broad range of commercial retail and residential uses, high quality streetscapes and a robust transit system" (City of Austin, 2009b, B4). The final element of the public process was the "visual translation workshop" where preferences found in the previous stage were translated into specific recommendations. Based on this analysis, a map was developed depicting the highest priority for redevelopment located in the western zone of the corridor – the location of the highest concentration of rental housing. In addition, a map of "public perceptions of existing conditions" was developed, showing participants'

---

6   In the first version of the plan, existing conditions were briefly described in an appendix, often through presentation of a single map, with little discussion and no analysis of the problems depicted. Conditions identified in the appendix were not linked to either discussion of plan goals nor of potential impacts of the proposed plan on existing conditions.

7   The plan describes a process in which a select group of early participants ("stakeholders") were extremely influential in setting the frame for later discussions. These early participants included "those who have been active in the planning or development process in the area, such as individuals who contributed to [plans for overlapping neighborhoods], business owners, landowners, developers, and other community interest groups." The focus of these discussions skewed heavily toward defining the parameters for redevelopment – something likely of material interest to the participating developers and landowners.

8   The 11 elements reviewed included streets, pedestrian realm, commercial, mixed-use and residential development, parking, signage, parks/plazas/open space, placemaking, sustainability, and mobility. Images were scored on a scale of -10 to $+10$, means and standard deviations were generated and the images scoring highest were the basis for recommendations in the plan.

views of crime hot spots, dangerous intersections, speeding zones, flood prone areas and viewsheds to protect.[9]

The first draft of the report was released (in English) publicly in June of 2009, and was distributed to libraries and posted on-line for comments. Its release coincided with discussions of a redevelopment project in the westernmost stretch of the corridor which, since approved, has now resulted in the demolition of approximately 500 units of low income (but unsubsidized) multifamily housing. Concerns over the lack of representation in the planning process among low-income immigrant renters became a focal point among affordable housing advocates. After a group of low income renters, all immigrant women, spoke about their concerns at a committee meeting, planning commissioners directed staff to present the proposed plan to the community in Spanish. The women's participation, along with that of housing advocates (including myself) resulted in substantial additions to the plan. Neighborhood demographics and more detailed information about affordable housing, local schools and other public facilities serving area residents were included in the revised plan, which provided a broader context for discussions of community benefits.

Yet, the participation of the immigrant women did not fundamentally change the substantive discussion related to planning goals and analyses. The maps created through the vision translation process described earlier, in which zones prioritized for redevelopment overlapped precisely with areas defined as important by the women immigrants, were still presented as valid. The final plan was approved by the planning commission in February 2010 and adopted by city council in March.

With its emphasis on attracting future residents and future outcomes related to ecological and economic neighborhood revitalization, the planning process and goals were not grounded in the current social and environmental contexts. And while there was some recognition of the ecological vulnerability of the neighborhood (i.e., risk of flooding and water quality of urban creeks), there was little meaningful recognition of existing vulnerable social groups, and none of the data indicating their various vulnerabilities were included. Indicators that have proven useful to disaster researchers trying to predict vulnerability following natural disasters include poverty, low household income, unstable employment, renter status, lack of household assets, lack of access to services (including schools and other community institutions), and strength of social networks. Given the East Austin context, linguistic isolation and legal status are also relevant. East Riverside is overwhelmingly low income, and populated by renters (85 percent of residents in 2000). In 2008, nearly a third of households had incomes below $15,000; 47 percent fell below $25,000 per year.[10]

---

9   Comparing the two maps, we find no correlation between any of the negative features on the existing conditions map (including crime hot spots) and redevelopment priorities.

10   While some of these households may represent university students, demographic trends suggest that the student population is a minority and continues to shrink.

A limited set of interviews carried out with residents and community leaders including school principals, ministers and flea market vendors revealed that the neighborhood is also home to a large community of Latino immigrants, mostly families. Many men work in various construction-related occupations, some as day laborers. Census data do not fully account for the presence of undocumented immigrant households. According to the 2000 Census, non-family households were overrepresented in the neighborhood, ranging from almost 60 percent to over 68 percent of all households in area census tracts. Similarly, only 11 percent of households were considered "linguistically isolated." Households with children formed only between 11.4 and 25.5 percent. Administrative data give us a fuller and distinctly different picture: the seven local elementary schools serving this neighborhood educate very low-income, Latino students, many of whom qualify for bilingual services (see Table 6.1).

East Riverside residents are also more likely to be transit dependent relative to other city neighborhoods. Two of the most heavily travelled bus lines in the city pass through the area; almost three times as many local residents commuted to work by bus as did citywide in 2000 (11 percent vs 4 percent). A disproportionate share of residents also carpool (19 percent vs 14 percent citywide) (US Census, 2000). Interviews with women revealed that low income families are especially transit dependent. The poor streetscape and lack of safe crossings produce high levels of accidents involving pedestrians.

Table 6.1 **Elementary schools serving East Riverside corridor, demographics, 2008**

Elementary school (state accountability rating)	Hispanic student population	Economically disadvantaged student population	Limited English proficiency population	Mobility rate 2006–07
Brooke (Recognized 2008)	87%	93%	29%	23.9%
Allison Elementary (academically acceptable)	90%	95%	38%	30.0%
Linder (academically acceptable)	90%	95%	60%	36.7%
Metz (Recognized 2008)	97%	93%	54%	25.3%
Sanchez (academically acceptable)	94%	92%	62%	30.8%
Baty (Del Valle ISD)	80%	90%	31%	30.0%
Travis Heights (unacceptable) 2008	68%	77%	24%	21.4%
District	57%	60%	24%	25.5%

Source: Demographic data and campus mobility rates are from the Texas Education Agency, Academic Excellence Indicator System, 2007–08 Campus Profile. Campus mobility rates are for 2006–07 (same source).

Low income households rely on a handful of services located in or near the neighborhood. These include a city health clinic, several private organizations assisting women with domestic violence, various youth programs, food assistance (WIC) and several programs offering services to people with disabilities. In addition, many area businesses cater to the immigrant community. Many small businesses, including a large weekend flea market whose vendors are also low income Latinos, would be adversely affected by redevelopment outcomes that failed to preserve retail space at affordable rents and failed to provide a space zoned for street vendors.

The East Riverside corridor includes ecological vulnerabilities that are tied to both its status as a low income community and its proximity to a dammed river and several creeks. Due to the large amounts of impervious surface associated with the historical concentration of apartment complexes and their large surface parking lots, water quality in the watershed within which most the ERCMP will be located is rated as poor by the City of Austin Watershed Protection and Development Review Agency (WPDRA). Due to the highly erosive quality of existing soils, combined with insufficient riparian buffers and development constructed directly in the riparian zone, sedimentation is a constant and serious problem. Habitat quality is also ranked poor. It is estimated that upgrading localized drainage systems within the Country Club Creek watershed will cost $13.3 million dollars, targeting 21 separate systems (City of Austin Watershed Protection and Development Review Agency, no date). The Austin Water Utility (AWU) concluded that existing water infrastructure (40–60 years old) is nearing the end of its effective utility. Within the proposed transit corridor there are no existing parks or open spaces (although there is a golf course adjacent to the zone). Degraded watershed functioning and the lack of park and open spaces both contribute to the ecologically degraded quality of East Riverside.

Historical analyses provide insight into the processes through which social and ecological conditions of vulnerability have been introduced, maintained and exacerbated. The processes associated with concentrating low income Latinos in this neighborhood are complex. However, it is possible to describe the factors leading to the concentration of apartments and subsidized housing in this zone and the ways that ecological vulnerabilities have been shaped by the same dynamics. Austin's tremendous population growth during the 1970s and 1980s was accompanied by a shift from small scale rental housing to larger, multiunit buildings. The 1979 Austin Comprehensive Plan emphasizes protecting older city neighborhoods from change due to growth pressures.[11] Multi-family (MF) residential areas were designated; however there is no reference to MF housing

---

11   "The quality of the city's housing and neighborhoods, including older, centrally located neighborhoods, is important to the well-being of every resident in the community" (Austin Tomorrow, 1979, p. 64). Stated sub-goals included: "Protect existing neighborhoods from the intrusion of higher intensity land uses" and "increase the power of neighborhood residents in decisions affecting the neighborhood" (Austin Tomorrow, 1979, p. 64).

in any of the housing goals in the 1979 plan. The lack of MF zoned land has been a chronic complaint of developers. According to the city demographer, 85 percent of MF projects put forward between January 1, 1992 and January 1, 1997 required a zoning variance – and 80 percent were granted their variance (Robinson, 2007). However, the pattern of concentration that resulted suggests that developers most often sought to locate developments away from established single family neighborhoods, thus avoiding likely opposition in the process of obtaining a zoning variance.

At the same time, the University of Texas was under increasing pressure to provide more housing for its students. By 1964, the University was housing only twelve percent of its student population. UT enrolled 27,345 students in 1967–68 and enrollment would reach close to 50,000 by 2000 (University of Texas, 2007). In 1969, the University began free shuttle bus service to enable students living in neighborhoods where apartments were concentrated to get to campus. Significant concentrations of student housing emerged along shuttle bus routes, including East Riverside. By 2007, the city's apartment stock had grown considerably, but was still concentrated in large complexes in neighborhoods with few single family residences (Austin Investor Interests, 2007). The East Riverside neighborhood is one such neighborhood. The neighborhood's disproportionate share of subsidized affordable housing is also part of a larger citywide pattern, through which areas with low land values (shaped by forces described above) and high poverty rates are the favored locations for multi-family housing.

In addition, the overconcentration of rental complexes in East Riverside is intimately connected to flooding and other environmental problems in the area. At the city level, Lee (2005) documented an 184 percent increase in residential development in floodplains between 1990 and 2000 (from 2,740 to 7,792 acres). During this same time period, the portion of low-income households living in floodplains rose from 31.3 percent to 43.7 percent (Lee, 2005, Table 8, 51), with the rate of increase of population in floodplains highest for low income households (83.4 percent vs 27.2 percent for the city overall). The 193 percent increase in floodplain acreage occupied by mobile homes was especially dramatic. By 2000, the share of mobile homes located in floodplains had risen from 7.5 percent to 12.9 percent (Lee, 2005, Table 5, 45). These patterns raise concerns that displacement of low income populations may result in their relocation to environmentally sensitive areas elsewhere in Austin.

Linking an analysis of the potential displacement of low-income households to ecological impacts produced in neighborhoods outside East Riverside provides insight into how displacement does not resolve problems but rather moves them. Historical analysis that focuses on the production of vulnerabilities facilitates an understanding of how patterns of investment, development and land use have contributed to the development of an ecologically and socially vulnerable neighborhood through neglect and inattention. The current approach to the ERMC plan appears to be based primarily on the shared interests of developers (who want to increase the intensity of development) and city officials (who want to increase the local tax base through development strategies). While city planners are leveraging the language of sustainability to frame the ERCMP, and include some of the "best practices" recommended by the APA or Smart Growth America, the lack of analysis about currently vulnerable populations challenges, and even contradicts, sustainability as a primary motivation.

Production of Urban Vulnerabilities as Ecological Gentrification

Both of these cases reflect how an ecologically focused agenda – as either green space land use designation or transit oriented development intended to reduce vehicle miles traveled on private transportation – assumes a moral authority for the decisions made in urban environments with a constricted notion of the public good. In each case, ideological notions of ecology intersect with notions of legitimate beneficiaries of ecological enhancement; these intersections contribute to processes that create, perpetuate and exacerbate conditions of vulnerability for already vulnerable human populations. In Seattle, the ecological benefits associated with setting aside large tracts of land are conveyed to the city and to the legitimately housed, while homeless people are made increasingly vulnerable to harm, arrest and banishment. Green space planning, with its commitment to urban ecological functioning, exacerbates existing harmful conditions and risks for the poorest people. The Seattle case represents the strategy of neglecting, and even exacerbating, a population's vulnerable condition in the service of specific ideologies about what constitutes an urban green space and who is granted access to this ecological resource. As an act of improving and enhancing urban ecological function – in terms of carbon sequestration, habitat connectivity and stormwater management – green space designation, in conjunction with enforcement of exclusionary policies, operate as a gentrifying process for the most vulnerable users. The benefits associated with green space designation are distributed unevenly, and displaced vulnerable people suffer material and juridical consequences.

Questions remain about the scale at which the designation of green spaces spurs ecological gentrification in Seattle. Are these designated green spaces spurring gentrification in the neighborhoods where they are located? Public trails, parks and green spaces have been found to increase property values (Crompton, 2001), potentially exerting pressure on economically stressed households who might be unable to afford rising property taxes. Long-term research might identify plausible cascading effects of these green spaces on shifting neighborhood demographics (wealthier residents moving in) and housing quality (improvement in housing stock over time).

In Austin, the multi-scalar productions of vulnerabilities – both social and ecological – are directly tied to strategies for neighborhood improvement. Reducing the threats of floods through alternative stormwater design strategies and increased access to light rail transit are highly desirable benefits that are predicted to drastically improve the quality of the neighborhood. However, the plan's failure to adequately address the preservation of affordable rental housing stock will initiate a gentrification process, resulting in the displacement of the poorest households seeking affordable rental housing on the urban fringe. The likely increase in VMTs associated with the necessity of acquiring private transportation among low-income immigrant households will most likely produce more carbon in the atmosphere, emit more greenhouse gases, and ultimately result in a larger carbon footprint. The co-production of social and ecological vulnerabilities, in this case, is evident only when the analytic focus shifts to the regional scale that links the urban and urban fringe.

Accurate quantification of the effects of displacing public transit-dependent low-income households to the urban fringe is important. These measures include numbers of displaced households, increases in private vehicle miles traveled, and other quality of life stressors resulting from displacement. Providing empirical evidence of how transit oriented development undermines the achievement of sustainability related goals and produces urban vulnerabilities for the region is a critical step towards potentially reforming the design and regulation of cities. Taken together, these future research efforts represent an urban justice agenda that highlights equity as a key component of the dialectics associated with sustainability (Dooling 2008, 2009).

Conclusion

These two cases demonstrate that social vulnerabilities can be produced from attempts to enhance ecological function; that processes related to ecological gentrification – in particular displacement through urban ecological improvements – are part of the driving forces behind the production of social vulnerabilities; and that, in the Austin case, corresponding ecological vulnerabilities emerge at a larger spatial scale.

The conventional concept of vulnerability, as a static point-in-time assessment of risk to a natural hazard, is re-conceived of as an urban political ecological phenomenon which develops over time and emerges from the contradictions associated with social, ecological and economic change. Ecological gentrification is one conceptual framework that highlights specific mechanisms through which multiple vulnerabilities are produced and experienced as states of being. Clearly, not all urban vulnerabilities are produced through ecological gentrification; not all experiences of urban vulnerability are related to threats of displacement and incarceration. While this may be framed as a conventional political economic analysis (the uneven distribution of benefits resulting from urban strategies driven by powerful political and economic elites), when situated in the context of sustainability discourse, ecological gentrification provokes a dialectical analysis assessing impacts associated with environmentally focused change.

The inconsistencies and contradictions in these two cases demonstrate the impotence of sustainability as a progressive concept in the face of urban political forces that take advantage of polarized perspectives of certain social groups.

Rather than sustainability maximizing the integration of economic, ecological and equity components in a diverse urban environment, these analyses show that environmental sustainability can be exploited as an urban planning ideology to establish tolerable levels of harm to the most vulnerable citizens in pursuit of differentiated ecological functioning.

References

Anderson, R. (1996). Homeless violence and the informal rules of street life. *Journal of Social Distress and the Homeless* 5(4), 369–380.

Arnold, K. (2004). *Homelessness, Citizenship, and Identity: The Uncanniness of Late Modernity.* Albany: State University Press of New York.

Austin Investor Interests (2007). Austin MSA Submarket Map 2007.

Bass, S. (2000). Negotiating change: community organizations and the politics of policing. *Urban Affairs Review* 36(2), 148–177.

Bawarshi, A., Dillon, G., Kelly, M., Rai, C., Silberstein, S., Stygall, G., Toft, A., English, T. and Thomas, B. (no date). *Media analysis of Homeless Encampment Sweeps.* Homeless Media Covereage Study Group. Seattle: University of Seattle.

Birch, E. and Wachter, S. (2008). *Growing Greener Cities: Urban Sustainability in the Twentieth First Century.* Philadelphia: University of Pennsylvania Press.

Bowden, C. (2010). *Murder City; Cuidad Juarez and the Global Economy's New Killing Fields.* New York: Nation Books.

Brookings Institute (2010). Metromonitor: tracking economic recession and recovery in America's 100 largest metropolitan areas. Retrieved from: <http://www.brookings.edu/~/media/Files/Programs/Metro/metro_monitor/2010_09_metro_monitor/0915_metro_monitor.pdf>.

Burt, M., Hedderson, J., Zweig, J., Ortiz, M., Aron-Turnham, L. and Johnson, S. (2004). *Strategies for Reducing Chronic Street Homelessness.* Washington, DC: US Department of Housing and Urban Development.

Campbell, H. (2009). *Drug War Zone: Frontline Dispatches from the Streets of El Paso and Juarez.* Austin: University of Texas Press.

City of Austin, Department of Planning and Development Review (2009b). East Riverside corridor master plan. Draft. June. Available from: <www.eastriversidecorridor.com> [Accessed 5 January 2010].

City of Austin (2009d). Downtown Austin plan – urban rail. Available from: <http://www.ci.austin.tx.us/downtown/dap_urban_rail.htm> [Accessed 20 January 2010].

City of Austin. (no date). Watershed Protection and Development Review Agency.

Cohen, A. (2008). Rent at an all-time high – if you can find a place. *Seattle Post Intelligencer.* Retrieved from the World Wide Web on April 24, 2008 from: <http://seattlepi.nwsource.com/local/359900_apartment21.html>.

Cohen, A. (2011). Area house prices post largest drops since 2009. *Seattle Post Intelligencer*. Retrieved from the World Wide Web on 3 March 2011 from: <http://www.seattlepi.com/local/436494_housing03.html>.

Crompton, J. (2001). The impact of parks on property values: a review of the empirical evidence. *Journal of Leisure Research* 33(1), 1–32.

Desjarlias, R. (1997). *Shelter Blues: Sanity and Selfhood Among the Homeless*. Philadelphia: University of Pennsylvania Press.

Dooling, S. (2008). Ecological gentrification: re-negotiating justice in the city. *Critical Planning*.

Dooling, S. (2009). Ecological gentrification: a research agenda exploring justice in the city. *International Journal of Urban and Regional Research* 33(3), 621–239.

Dooling, S., Simon, G. and Yocom, K. (2006). Place-based urban ecology: a century of park planning in Seattle. *Urban Ecosystems* 9(4), 299–321.

Ewing, R., Bartholomew, K., Winkelman, S., Walters, J. and Chen, D. (2008). *Growing Cooler: The Evidence on Urban Development and Climate Change*. The Urban Land Institute.

Farrell, M. (2010). Immigration reforrm is critical to economic growth in Texas. Retrieved from the World Wide Web on 1 October 2010 from: <http://blogs.forbes.com/maureenfarrell/2010/09/22/immigration-reform-is-critical-to-economic-growth-in-texas/>.

Fears, D. (2009). Shelter is even more tenuous: city budget cut stun advocates for homeless. *Washington Post*.

Feldman, L. (2004). *Citizens Withtout Shelter: Homelessness, Democracy, and Political Exclusion*. Ithaca and London: Cornell University Press.

Forman, R. (2008). *Urban Regions: Ecology and Planning Beyond the City*. Cambridge: Cambridge University Press.

Gibson, T. (2004). *Securing the Spectacular City: The Politics of Revitalization and Homelessness in Downtown Seattle*. New York: Lexington Books.

Gregor, K. (2010). Lucia Athens named chief sustainability officer. *The Austin Chronicle*. Retrieved from the World Wide Web on 3 September 2010 from: <http://www.austinchronicle.com/dev/blogs/news/2010-06-18/lucia-athens-named-chief-sustainability-officer/>.

Groth, P. (1994). *Living Downtown: A History of Residential Hotels in the United States*. Berkeley: University of California Press.

Hopper, K. (2003). *Reckoning with Homelessness*. Ithaca and London: Cornell University Press.

Klingle, M. (2010). Salmon and the Persistence of History in Urban Environmental Politics. In: C. Miller (ed.). *Cities and Nature in the American West*. Reno: The University of Nevada Press.

Koegel, P., Burnam, M. and Baumohl, J. (1996). Chapter 3: The causes of homelessness. In: J. Baumohl (ed.) *Homelessness in America*. Pheonix: Oryx Press.

Lee, D. (2005). The growth of low-income population in floodplain: a study of Austin, TX. Professional report (M.S. in community and regional planning) – University of Texas at Austin.

Liberto, J. (2011). Budget cuts may hit homeless vets. Retrieved from the World Wide Web on March 1, 2001 from: <http://money.cnn.com/2011/03/01/news/economy/homeless_veterans_housing_cuts/index.htm>.

Marin, P. (1987). Helping and hating the homeless. *Harper's Magazine*, 310–318.

May, J. (2000). Of nomads and vagrants: homelessness and narratives of home as place. *Environment and Planning D: Society and Space* 18, 737–759.

Mitchell, D. (2003). *The Right to the City: Social Justice and the Fight for Public Space*. London: Guilford Press.

National low income housing coalition. <http://www.nlihc.org/oor/oor2008/data.cfm?getstate=on&getmsa=on&msa=1186&getcounty=on&county=2936&state=WA>.

Newman, P. and Kenworthy, J. (2003). Sustainability and cities: summary and conclusions. In: A. Cuthbert (ed.) *Designing Cities: Critical Readings in Urban Deisng.* Malden, MA: Blackwell Publishing, pp. 235–242.

Ng, M., Grant, C., Donoso, R.E. and Eaton, E. (2007). Preserving affordable apartments in Austin: case study of the East Riverside/Oltorf combined neighborhood planning area. Paper prepared for CRP 388 Affordable Housing Policy.

O'Connell, J. (2005). *Premature Mortality in Homeless Populations*. National health care for the homeless council.

Portney, K. (2003). *Taking Sustainable Cities Serioulsy: Economic Development, the Environment, and Quality of Life in American Cities*. Cambridge: MIT Press.

Quastel, N. (2009). Political ecologies of gentrification. *Urban Geography* 30(7) 694–725.

Robinson, L., Newell, J. and Marzluff, J. (2005). Twenty-five years of sprawl in the Seattle region: growth management responses and implications for conservation. *Landscape and Urban Planning* 71, 51–72.

Roper, R. (1988). *The Invisible Homeless: A New Urban Ecology.* New York: Insight Books.

Rossi, P. (1989). *Down and Out in America: The Origins of Homelessness.* Chicago: University of Chicago Press.

Rowe, S. and Wolch, J. (1990). Social networks in time and space: homeless women in skid row, Los Angeles. *Annals of the Association of American Geographers* 80(2), 184–205.

Schrag, P. (2010). *Not Fit For Our Society: Immigration and Nativism in America.* Berkeley, CA: University of California Press.

Scott, S. (2007). *All Our Sisters: Stories of Homeless Women in Canada.* Ontario: Broadview Press.

Seattle Parks and Recreation Department. (No Date). West Duwamish greenbelt parks and green spaces levy project information. Retrieved from the World Wide Web on 3 January 2011 from: <http://www.cityofseattle.net/parks/projects/west_duwamish/greenbelt.htm>.

Seattle Parks and Recreation (2000). Seattle parks and recreation plan 2000: an update to the 1993 parks COMPLAN. Seattle, Washington, p. 86.

Shinn, M. and Gillespie, C. (1994). The roles of housing and poverty in the origins of homelessness. *American Behavioral Scientist* 37(4), 505–521.

Shrestha, B. (2007). Median house price in Seattle tops $5000,0000. *Seattle Times*.

Streitmatter, R. (1999). The nativist press: demonizing the American immigrant. *Journalism and Mass Communication Quarterly* 74(4), 673–683.

Swearingen, W. (2010). *Environmental City: People, Place, Politics and the Meaning of Modern Austin*. Austin: University of Texas Press.

Turnbull, L. (2009). Report disputes immigrants' drain on state and local economies. *The Seattle Times*.

United States Conferences of Mayors (2008). Hunger and homelessness survey: a status report on hunger and homelessness in America's cities. Washington, D. C.

United Way, City of Seattle, Committee to end homelessness King County (2009). Seattle homeless needs assessment.

University of Texas. Division of Housing and Food (2007). An overview of housing at the University of Texas. Available from: <http://www.utexas.edu/student/housing/index.php?site=0&scode=0&id=725> [Accessed 7 October 2009].

US Census (2000). SF3. P30. Means of transport to work for workers 16 years and over. For zip code 78741.

Waldron, J. (1991). Homelessness and the issue of freedom. *UCLA Law Review* 39, 295–324.

Western Regional Advocacy Regional Project (2006). *Without Housing: Decades of Federal Housing Cutbacks, Massive Homelessness, and Policy Failures*. San Francisco.

Zukin, S. (1995). *The Cultures of Cities*. Cambridge: Basil Blackwell.

Chapter 7

Between Here and There: Mobilizing Urban Vulnerabilities in Climate Camps and Transition Towns

Kelvin Mason and Mark Whitehead

Introduction: Grow Heathrow and In-between Vulnerabilities

Our story begins on a small piece of derelict land in the heart of Sipson, West London. Sipson is a classic example of what Sieverts (2003) describes as an *in-between* city community. Sandwiched between the M4 motorway to the north, and Heathrow Airport to the south, Sipson is a semi-rural area surrounded by the transport infrastructures that enable the city of London to function. It is an in-between urban space because it comprises an uneasy blend of the edge-city and green-fringe. On the first of March 2010 Sipson witnessed what is known in the parlance of guerrilla gardeners and climate activists as a "site-take." The site-take involved around twenty environmental activists occupying the aforementioned derelict land in order to convert it into a community garden and environmental education centre. The location, home to a series of dilapidated greenhouses, was routinely used for car-breaking and illegal waste tipping. In clearing the site, activists had to remove thirty tonnes of waste before work could start on the creation of a community garden. Perhaps the lack of care shown to the site should come as little surprise given that Sipson was to be located at the very centre of the planned Heathrow Airport expansion, and effectively cemented over by the new terminal building and runway.

Grow Heathrow, as the community garden is now known, actually represents a compelling geographical convergence point for two of the most prominent environmental movements in Britain today. Grow Heathrow has been primarily inspired by the principles of the Transition Culture movement (Hopkins, 2008a). The Transition Culture movement is a grassroots initiative that focuses on the development of locally centred community responses to the twin threats of climate change and peak oil production (Mason and Whitehead, *forthcoming*). The fostering of local food production and environmental re-skilling at *Grow Heathrow* are hallmarks of the eco-empowerment and green localism that characterise Transition Culture. The second movement that undergirds *Grow Heathrow* is the Camp for Climate Action. The Camp for Climate Action is a British activist network that strategically occupies prominent political and economic locations (including power stations, the headquarters of banks, and G20 summits) in order to bring attention to the lack of progress on addressing the threat of climate change. Although climate camps support similar eco-pedagogic initiatives as those evident in Transition Culture projects, both offering a space within which people can learn about sustainable living techniques, they are far more transitory and politically confrontational than those spaces associated with the Transition Culture movement.

Grow Heathrow is a hybrid manifestation of the Transition Culture movement and the Camp for Climate Action to the extent that it attempts to develop permanent community based solutions to ensuing urban-based environmental threats (Transition Culture), while "illegally" occupying a site that signals direct political opposition to the industrial–governmental complex's inaction on climate change issues (Camp for Climate Change). In addition to embodying a fascinating and contentious political amalgam (Trapese Collective, 2008; Mason, 2008; Hopkins, 2008b; Hopkins, 2008c), however, *Grow Heathrow* also provides a point of departure for discussion of the political construction and mobilization of urban vulnerabilities in Britain today. In different (but sometimes overlapping) ways, the Transition Culture movement and the Camp for Climate Action illustrate three issues concerning urban-environmental vulnerabilities:

1. the different ways in which vulnerabilities are socially constructed;
2. differential political mobilizations and uses of vulnerability; and
3. the connections that exist between spatial politics and struggles over vulnerability.

A central aspect of this chapter is a realization that many of the most pressing environmental threats facing cities (particularly climate change and peak oil) embody and produce spatially uneven forms of vulnerability. Consequently, while large-scale environmental threats embody challenges to particular urban communities in aggregate, they also constitute a complex moral geography. This moral geography operates at a local level in relation to the ways in which different segments of a given urban population are seen to be more vulnerable than others to impending environmental crises. At a larger spatial scale, this moral geography sees some cities being more vulnerable (whether because of location or institutional capacity) to environmental threats as a whole than others. This maelstrom of ethical dilemmas is further complicated by the highly differentiated levels of responsibility that different urban communities around the world have for the production of contemporary environmental threats. In exploring questions of urban-environmental vulnerability, we use this chapter to raise important questions concerning the impact that different articulations of metropolitan vulnerability have on the ethical constitution and empathies of urban communities. Through the different actions of the Transition Culture movement and the Camp for Climate Action, we consider how approaches to urban-environmental vulnerability raise important questions about our responsibilities to urban neighbours and to those people who we may never meet, but whose life chances are shaped by contemporary urban life. This is an analysis of urban vulnerability that combines a concern for the proximate urban here, and more distant metropolitan there.

This chapter begins by offering some provisional observations about the nature of urban environmental vulnerability, which draws specific attention to how the mobilization of a politics of vulnerability can lead in very different normative directions and connect to different forms of spatial politics. The second section of the chapter considers the construction of a particularly local sense of urban vulnerability within the Transition Culture movement, and how this leads to a potentially closed urban ethic of care. The third section proceeds to consider the more embodied and distantiated (that is, spread over space) forms of vulnerability that get expressed in the Camp for Climate Action. These two case studies are utilize to identify three modalities of urban vulnerability that often exist alongside each other within the metropolitan experience: 1) externally produced vulnerabilities (visited on cities by forces that operate at scales above and beyond the city); 2) internally constructed metro-vulnerabilities (or those that are constructed by different groups within a city to serve particular political ends); and 3) experiential vulnerabilities (namely the physical experience of being vulnerable by residents within a city either as a product of external threats, or because of their participation in the internal construction of a politics of vulnerability). Drawing on the preceding reflections on Transition Culture and Climate Camps, the concluding section of this chapter explores the potential for a convergent politics of urban environmental vulnerability that lies somewhere in between localism and internationalism; care at a distance and long-term commitments to specific places; radicalism and safety. This is a kind of politics of urban vulnerability that is able to position the city in relation to the geo-economic drivers of risk, of which it is undoubtedly a part, but also embrace the city as a highly creative locus for the production of sustainable forms of environmental security (see Merrifield, 2002).

On the Nature and Spatial Politics of Urban-environmental Vulnerability

Vulnerability as Political Project

Accepting a definition of urban vulnerability as the susceptibility of urban dwellers to physical or emotional injury, we are interested in how social movements view and mobilize this susceptibility. First, if urban denizens are vulnerable, there must be the threat of harm and so we must be interested in the nature of that threat, and indeed of that harm. In terms of climate change and peak oil, the threat of harm appears to be something that comes to the city from outside and is beyond the ability of an urban community to control. This is what we term an *external vulnerability*: those vulnerabilities that a city is deeply implicated in the production of, but which are part of processes of formation and construction that transcend the city. Regardless of the scale at which vulnerability is produced, it must also undergo a process of social construction upon entering the urban political arena. *Internally constructed* forms of vulnerability involve the appropriation of existing narratives of vulnerability (such as those surrounding peak oil and climate change), and their translation into meaningful statements of threat for a given urban space. This internalization process is of high political significance. If these constructed threats are framed as assaults on the city, that is on the physical or emotional health of urban dwellers, then urban movements seeking to safeguard cities from harm might construct the threat as *fait accompli* and concentrate their efforts on developing adaptation measures, for example by building flood defences. Other, more radical responses, might seek to stop the flood from happening. Politically, such mitigation may mean, seeking out and challenging those responsible for the continuing assault. Though they are certainly not definitive, it may be useful to frame a tentative view of such conceptions of adaptation and mitigation as broadly representing the responses to a threat of, respectively, the Transition Culture movement and the Camp for Climate Action. Both responses are rational and may depend principally on a movements' analysis of the relational nature of the threat (see Massey, 1992, 1994, 2007). But what this broad spectrum of political response to urban environmental threat reveals are the varying political opportunities that vulnerability presents to metropolitan actors.

In *The Shock Doctrine* Naomi Klein illustrates how neo-liberal capitalism and free-market economic thinking has exploited extended moments of vulnerability to re-conceive social space as radically privatised, a pattern she brands (*sic*) "Disaster Capitalism" (Klein, 2008). On these terms, the political opportunities presented by intense moments of vulnerability (following, for example, the events of Hurricane Katrina) are not so much questions of how to address vulnerability, but how to utilize the socio-psychology mêlée that often follows a disaster to pursue a range of ideologically inspired political projects. On the other hand, Rebecca Solnit highlights the possibility for community that arises in those moments of vulnerability that follow disaster (Solnit, 2009).[1] In those moments, where conceived space is disrupted and social practices must change, lived space becomes a critique of society: dismal privatisation is undone and citizens are returned to a meaningful collective life and experience a paradoxical joy. Solnit proposes that in these moments the repressed imagination of a better world is brought to life. It is, she claims, a glimpse of paradise from the jaws of hell. As Mike Davis reminds us, however, the moment is likely to be all too short as the dismal, joyless and disconnected is swiftly reconceived and reconstructed by the forces of capital, greedy and corrupt, sowing seeds in the very material fabric of the city that will ensure the next disaster is compounded, that tragedy is assured within the logics of *seismic Keynesianism's* disaster accumulation (Davis, 2000). What the work of Klein, Solnit, and Davis inter alia, emphasise fundamentally is the experiential dimensions of vulnerability. Experiential vulnerabilities relate to

---

1   Solnit also warns that the agencies of the state are unlikely to recognize or assist suddenly autonomous and hence unfamiliar and perhaps threatening communities.

what actually happens to individuals and communities who become vulnerable. Existential encounters with vulnerability run along two political axes. First, exposures to vulnerability produce the opportunities to pursue very different political visions in response to threats (communitarian or neo-liberal for example). However, an often-overlooked relation between the experience of vulnerability and politics is that political movements that seek to draw attention to future threats can actually place their members in positions of vulnerability in the shorter terms themselves (see discussion on the Camp for Climate Change later).

In this chapter we thus perceive of the different modalities of urban-environmental vulnerability as providing the momentary opportunity for a range of more or less radical, more or less conservative, and more or less progressive political mobilizations around the metropolitan future. Crucially, however, we assert that these coalitions of vulnerability are not simply political movements of the city, but are bound-up within the deeper spatial politics of metropolitan life.

Urban Spatialities and the Condition of Vulnerability

In order to understand the nature of the political movements that emerge from conditions of urban vulnerability we find the work of Henri Lefebvre and Doreen Massey to be of great utility. What both Lefebvre and Massey help to illustrate, in different ways, are the connections that exist between the spatial constitution of the urban and the different types of political practice that emerge within the urban arena. Such sensitivity to the spatial dynamics of urban politics helps to reveal that the struggle over urban-environmental vulnerability is more than merely a struggle for different visions of security: it is struggle over the very nature of the urban condition itself.

Table 7.1   Modalities of urban vulnerability associated with climate change and peak oil

Mode of Vulnerability	Key Features
External	Perceived as originating within socio-economic and environmental systems that are connected to the urban system, but are beyond a given cities unilateral control (including, global economic recession, pandemics, energy crises, near future climate change scenarios *inter alia*).
Internally Constructed	The translation of externally constituted vulnerabilities into meaningful and specific forms of threat to cities (for example the potential impacts of climate-induced flooding and inundation for coastal cites).
Experiential	The experience of urban vulnerability within the everyday lives of urban denizens. This form of vulnerability can relate either to existential encounters with impending socio-ecological threats, or the physical experience of the vulnerabilities that are created during periods of socio-ecological turmoil.

For Lefebvre, space is not passive, not a surface upon which activities are reproduced (Lefebvre, 1991). Rather, space is itself produced, and as such it is a player in reproducing social life, or indeed producing it differently. Participation in the production of space holds emancipatory potential, the chance to disrupt the reproduction of the spectacle of capitalism. Lefebvre conceptualizes space as a "spatial triad" or "triple dialectic." The three "moments" he identifies are: *representations of space*, which are always conceived in the abstract, are constructed by professionals such as architects, planners, and engineers, and which inevitably have power and ideology embedded in the representation; *spatial practices,* perceived material spaces which structure everyday reality, ensuring social cohesion and continuity, for example the spaces that link home to job (the street, bus queue, bus journey, arrival at work); *representational spaces*, lived spaces of daily experience, spaces of emotion and instinct, contingent and creative spaces wherein, most significantly, representations of space and spatial practices can be imagined differently.

We suggest that filtering the emerging politics of urban–environmental vulnerability through the lens of Lefebvre's spatial *trialectic* provides helpful insights into the operational dynamics of these movements. At one level, *representations of vulnerable urban space* are promoted by a new breed of ecological metropolitan modernizers who operate under the discourses of sustainable urbanism, climate security, and the green city (what we term *internally constructed vulnerabilities*). The official construction of cities as environmentally vulnerable is paving the way for new regimes of urban financial accumulation and investment which seek to secure the city through the lucrative allocation of new infrastructural projects and green architectures (Hodson and Marvin, 2009). Far from empowering local people to help to develop the necessary social capital to support the implementation of resilient infrastructures (see Comfort, 2007), such visions of urban environmental security appear to be mainly about empowering elites who are devoted to securing and preserving a city's global competitive edge (*ibid*). On the other hand, it is clear that contemporary *representational spaces of vulnerability* are spurning new, creative, and seemingly empowering forms of urban politics (what we term *experiential vulnerabilities*). For Human Geographers and other social scientists, representational or lived spaces are where progressive new ideas about society and the future can be produced, spaces from where we might "reclaim, for its citizens, the space of our cities" (Merrifield, 2002, p. 181). It is, of course, precisely in this representational realm that *spatial practices* which cause and sustain the production of vulnerabilities are most easily identified and articulated. It is precisely in this context from the slums of Mumbai and Mombasa to the reclaimed streets of Portland, that urban community politics are emerging which both challenge the logic of dominant representations of urban vulnerability and identify how emerging socio-environmental threats can be tackled. These are political communities who do not just seek the right to urban environmental security, but the right to determine how this security should be achieved and what type of urban experience is left at the end of this period of securitization (see Gibson-Graham, 2003).

If Lefebvre's analysis of the spatial nature of urban politics enables us to understand the different geographical contexts within which responses to vulnerability emerge, it provides far less insight into the impact of vulnerability on how the spatial relations that sustain a place are perceived and politicised. What is clear, however, is that whether it comes from the representations of urban planning, or the representational registers of the street, vulnerability can often lead to the cultivation of a *closed city*: a place looking after its own interests in an era of environmental austerity and crisis (see Hodson and Marvin, 2009). In trying to think beyond the politico–ethical norms of the closed-city, Massey's work on the relationality of urban space is particularly instructive to contemporary work on vulnerability. In *World City,* Doreen Massey poses the ethical question: "What does this place stand for?" (Massey, 2007). This is a relational question and might be extended by urban movements responding to threat and vulnerability into the questions: "What does this movement stand for and (so) whom do we stand with, and how should we act?" Massey argues against localism, identifying the need to construct a politics that "thinks beyond the local […] If the reproduction of life in a place, from its most spectacular manifestation to its daily mundanities, is dependent upon poverty, say, or the denial of human rights elsewhere, then should (or *how* should) a 'local' politics confront this?" (Massey, 2007, p. 15). Massey thus proposes a politics of responsibility that would surely recognize that the securing of some places against the threats of vulnerability may actually make other places more, not less, vulnerable.

Perhaps the key point for mobilising urban vulnerabilities that Massey highlights is the strategic advantage – the potential strength – in place-based struggles. Elsewhere she highlights this neglected potential:

> There is an overwhelming tendency both in academic and political literature, and other forms of discourse, and in political practice to imagine the local as the product of the global but to neglect the counterpoint to this: the local construction of the global […] On this view local places are not simply always the victims of the global; nor are they always politically defensible redoubts *against* the global. Understanding space as the constant open production of topologies of power points to the fact that that different "places" will stand in contrasting relations to the global. They are differently located within the wider power geometries (Massey, 2004, p. 101).

Massey highlights for us the different locations of places in wider power geometries and so flags up strategic responses that are sensible of this and sensitive to it. In one instance, place-based struggle may be an effective tactic (that is, emphasising internal vulnerabilities); in another an active politics that transcends place and produces lived, translocal solidarities will be most appropriate (that is, mixing an external and internalizing perspective on vulnerabilities). The mobilisation of a politics of vulnerability can, then, lead not only to very different normative directions, proximate care versus global responsibility, for example, but also to different strategic responses (see Kaika, 2003). At any moment in space–time, a strategic response may mask a particular ethics of care, politics of responsibility or commitment to place.

Urban Transitions and Super-wicked Vulnerabilities

Transition Cultures and Preparatory Urbanism

A 2009 survey identified 94 transition communities in the UK and a further forty worldwide which had formally registered with the Transition Network, the constituted organising hub of that movement (Seyfang, 2009, p. 2). The majority of these communities are located in towns and cities. Since the inception of the groundbreaking Transition Town Totnes in 2005, the rapid rise of the Transition Culture Movement has been a phenomenon in British society.

The Transition Culture initiatives that operate in urban space represent an interesting example of preparatory, or anticipatory, urbanism (Hodson and Marvin, 2009). Preparatory urbanism is diametrically opposed to both urban modernity, which pursues the exhilarating mission of unfettered metropolitan transformation and growth, and sustainable urbanism, which embraces a more cautiously optimistic vision of carefully managed urban development (see Whitehead, forthcoming). Preparatory urbanisms take an altogether more sober and pessimistic view of the urban future. As a kind of urban response to the post-9/11 world, preparatory urbanisms suggests that we should prepare for metropolitan futures that are marked by severe shocks and ruptures to the everyday structures of urban life (see Davis, 1999). Some preparatory forms of urbanism take environmental issues as their primary focus. On these terms, issues of urban environmental preparation tend to take two broad forms: 1) preparation for the effects of shortages in key environmental resources (including, *inter alia*, food, water, energy); and 2) preparation for the socio-economic effects of near future environmental changes (including, *inter alia*, flooding, firestorms, hurricanes, and the spread of diseases).

Two things are particularly interesting about Transition Cultures as a form of preparatory urbanism. First, and unlike many of the preparatory forms of urbanism charted by Hodson and Marvin (2009), Transition Cultures are not orchestrated by coalitions of urban political and corporate elites in the pursuit of the competitive socio-economic benefits that can come from "securing" the necessary environmental resources and infrastructural protections that continued economic expansion requires. Rather, Transition urbanisms are community-based mobilizations that focus on the use and development of the latent talent, ability and general nous of urbanites to deal with environmental problems that may confront cities in the future. Second, Transition urbanism takes its primary environmental preparatory issues to be both peak oil and climate change. This twin focus marks

Transition Culture out in relation to both the forms of the urban vulnerability it seeks to expose, and the nature of the solutions that it proposes to these threats.

Peak oil production and climate change both embody what has been described as *super-wicked* environmental problems (Lazarus, 2009). Wicked environmental problems are those that are difficult or impossible to solve. Super-wicked environmental problems are, however, marked-out by a level of global socio-economic complexity that not only means there are no obvious technological solutions to a given environmental problem, but that even if there were, prevailing cultures, political institutions and geographical structures of life would make their implementation difficult to achieve (*ibid*). Lazarus also claims that super-wicked environmental problems are issues whose resolution actually become more difficult and costly as time goes on. If we can agree that *taken individually* peak oil production and climate change both deserve the moniker "super-wicked," attempts to tackle them as a unitary whole leave one scrambling for suitable superlatives (*sic*) to describe this condition. It is important to consider precisely what Transition urbanism proposes we do to tackle these twin threats.

At a very basic level, urban-based Transition Culture movements draw specific attention to the connections between the intertwined crises of peak oil and climate change, and the form and nature of cities. In relation to peak oil, Transition Culture movements emphasize that the spatial growth and functional segregation of cities has been facilitated by readily available cheap oil. The suburban spread of cities, and the separation of work, home, and leisure and consumption that this routinely involves, would have been unimaginable without the automobile (see Molotch, 1976; Wolch et al., 2004). As a political movement, Transition Culture invites us to consider the impact which increasingly scarce petrochemical resources will have on the spatial functioning and structured coherence of contemporary urban regions. In addition to disrupting the daily cycles of life at an urban scale, however, Transition Cultures also emphasize the destabilizing impacts that peak oil production will have on the global connections of trade and migration that currently define the globalized economies of so many cities (see Massey, 2008). What is perhaps most significant about Transition Culture is that it does not utilize the threat of peak oil to support an eschatological narrative of the future city (see later). As a classic form of preparatory urbanism, Transition Culture argues that the consequences of peak oil can be anticipated and effectively addressed. What is thus proposed is the heading-off of a full-on metropolitan energy crisis through a process of planned *energy descent* (Hopkins, 2008a; Hopkins and Lipman, 2009). Planned energy descent involves a careful assessment of how an urban community's energy budget can be reduced in a safe, socially just and sustainable way. At the heart of the energy descent philosophy is that organising social and economic life at urban-regional and global scales will be unsustainable in an energy lean future, and that re-localizing urban economies around much smaller metropolitan communities should be a key planning objective (Bailey and Hopkins, 2010; North, 2010). In addition to prioritizing the increasing localised production of a range of goods and services, the re-localization agenda pursued within Transition urbanisms also emphasizes the importance of empowering local communities to share and develop the skills that they have in order to meet their own needs.

If peak oil is seen as a threat to the spatial form and functioning of cities, the Transition Culture movement claims that climate change problematises the spatial location of urban communities. Essentially if the spatial sprawl of cities has been facilitated by cheap oil prices, relatively stable climatic conditions over the last 10,000 years have enabled urban communities to occupy a wide range of locations. Significant forms of future climate change, it is argued, could challenge the ability of some urban communities to retain access to secure sources of water and food required to sustain themselves, while other urban areas may become much more susceptible to the threats of flooding and hurricane damage. Key goals of urban transition are thus to carefully plan how alternative local food and water supply systems could be put in place and how communities could work together to both prevent climate related urban threats from materializing (or how to manage such problems should they occur) (cf. Newman, 2008 et al.). Of course, there are key areas of overlap between Transition Culture's concern with the impact of climate change and peak oil on cities. The threat of climate change may necessitate urban energy descents before society reaches a peak oil production threshold. At the same time, the impacts of climate change on food production regions may make depending on a global supply line of food both an increasingly expensive, and also unreliable urban planning option.

The key concept that unifies Transitions Culture's response to the twin crises of peak oil and climate change is that of *resilience*. While notions of social, economic, and ecological resilience are currently utilised in a range of different contexts to justify different forms of urban policy, in a Transition context resilience is taken to denote a very specific course of action (see Comfort, 2007). Drawing on the forms of resilience that build up in ecological systems, Transition urbanism encourages *diversity* (in food supply, energy production; infrastructure provision); modularity (or the creation of systems of supply and circulation that can operate independently of each other); and tightness of *feedback loops* (to ensure that if the systems that help to sustain a community are coming under threat, rapid remedial action can be take to address them) (see Hopkins, 2008a, p. 55). At the heart of the systems of urban resilience promoted by Transition Culture is recognition that metropolitan communities are going to face significant shocks to their prevailing modes of operation, and that building strategies and systems that can successfully absorb changes in these processes should be key priority of urban policy. We will return to the concept of resilience.

The Vulnerabilities of Transition

At one level, the political mobilization of notions of socio-ecological vulnerability may seem a startlingly obvious part of the Transition Culture movement. Urban society is facing two converging geo-historical threats of unprecedented proportions, and the very nature of metropolitan life is to become exposed to structural forms of vulnerability. But looking more closely at the movement we see a much more complex pattern of constructing and mobilizing vulnerability emerge; revealing important insights into the politics of vulnerabilities manifested in urban centres.

The types of vulnerability being framed and identified within Transition cultures take two basic forms: (1) recognition of an ontological socio-ecological threat to urban ways of life, and (2) a concern with the forms of responses that may emerge to that threat. Let us consider the issue of the ontological socio-ecological threat first. What is interesting about Transition Culture is that notions of vulnerability are not constructed as something which visit an urban community as an outside force (that is, world oil market prices, or sea level change), but as an outcome of inappropriate scale at which urban life is constructed (this is particularly evident in relation to peak oil, but also applies indirectly to issues of climate change). Vulnerability is thus cast at the intersection between the ontological threats of peak oil and climate change, and the nature of the spatial relations (both within and beyond a city) upon which an urban community is based. This way of constructing vulnerability is politically significant because it provides a localising pivot-point around which urban communities can begin to see the power that they have to insulate themselves against the emerging threats to their way of life. This localizing ethos of vulnerability does, however, tend towards a proximate ethic of care within which local communities primarily become concerned with the impacts of peak oil and climate change on their communities and what they can do about it, rather than considering the potential action that their communities could take to help those peoples who are most vulnerable to these twin threats in global society (see Mason and Whitehead, forthcoming). There is, it is important to note, awareness within Transition Culture communities of the differential vulnerabilities that peak oil and climate change present to different social groups (with particular concerns expressed for low income households and the elderly). The point is that this sense of differentiated vulnerability tends to operate predominately in relation to social groupings within a locality, and not geographically in relation to the highly uneven exposures to risk that some cities experience compared to others.

The second politics of vulnerability that is evident within Transition communities relates to the concerns over the socio-cultural response to questions of vulnerability itself. Addressing the question of how people respond to revelations of their impending vulnerability leads us into interesting questions concerning the psychology and *neuropolitics* of experiential forms of vulnerability. The psychological concerns of Transition Culture are captured nicely by movement guru Rob Hopkins' evocative notion of "post-petroleum stress disorder" (2008a, p. 80). Behind this medical metaphor lies a real concern of Transition Culture: that within the revelation of vulnerability there is a danger of deepening and perpetuating the very forms of vulnerability you seek to eradicate. As many people, including ourselves, who have worked within the Transition Culture movement can attest, presenting people with the convergent crises of peak oil and climate change can illicit a range of potentially harmful responses. These can range from the more obvious responses (including denial), to ones that are more harmful to the individual concerned (fear and paranoia) (see Hopkins, 2008a, pp. 81–82). The least desired responses to vulnerability for Transition urbanists would be recourse to *survivalist* tendencies or *external solutions* (*ibid*). Survivalist impulses are pernicious to Transition Culture because they suggest the need for individual self-reliance and social disintegration precisely when, Transitioners would argue, communities need to come together to share skills and develop systems of co-intelligence that can enable local communities to become more self-sustaining. According to the principles of Transition Culture, the recourse to *external solutions* (whether that be new technological breakthroughs, or the actions of local and national government) only tends to add to local vulnerabilities as the rapid experimentation, development, and implementation of local strategies for local resilience are delayed by the wait for external interventions that never come (*ibid*). Given that the Hirsch Report on the *Peaking of World Oil Production: Impacts, Mitigation, and Risk Management* (Hirsch et al., 2005) suggested that it may take between ten and 20 years to make the transition away from a petroleum-based economy, it is clear how such delays could actively contribute to the escalation of vulnerability.

In terms of the social psychology of vulnerability, what Transition Culture illustrates most clearly is the importance of utilising the potentially disempowering affects of feeling vulnerable as a basis for generating optimism (see Homer-Dixon, 2003). To these ends Transition Culture has focused less attention on the threats driving the need to transition, in order to emphasize the socio-cultural benefits of transition itself. Transition Culture has thus rapidly become associated with a slower pace of life, high levels of community cohesion and social care, and the happiness that can often result from being in control of the systems that make your day-to-day life possible. What such mobilizations inevitably lack is a critical take on the economic and political forces that drive the production of vulnerability and actively disempower many communities from adequately defending themselves against such threats.

Having suggested how vulnerabilities are framed by Transition Culture, let us revisit the concept of resilience. The Transition Culture movement develops its hybrid conception of resilience from understandings developing within the sciences of ecology and engineeringt. The Resilience Alliance, a multi-disciplinary research organization, endorses an engineering inspired interpretation of resilience as "the capacity of a system to absorb disturbance, undergo change and still retain essentially the same function, structure, identity, and feedbacks" (Resilience Alliance, 2010). In ecological terms, however, resilience refers to a systems moving from one energy-based regime into a new register of operation. In psychology, though resilience may be articulated as the ability of a person to return to a previous (original) state, it is viewed not as an individual attribute but rather as a (social) process. The concept of resilience developed in *The Transition Handbook* "refers to their ability to not collapse at first sight of oil or food shortages, and to their ability to respond with adaptability to disturbance

... Increased resilience and a stronger local economy do not mean that we put a fence up around our towns and cities and refuse to allow anything in or out [...] What it does mean is being more prepared for a leaner future, more self-reliant, and prioritising the local over the imported" (Hopkins, 2008, pp. 54–55). So in the Transition Movement resilience is interpreted as between optimistic engineering thinking and more pessimistic ecological models. It is not about bouncing back to an original state, or falling dramatically into a new entropic condition: it becomes about the ability to adapt and to change while retaining essential functions (see Gunderson and Holling, 2002), all framed within a local, though not isolated, system. Transition Culture resilience is not the antonym of vulnerability but rather a (preparatory) response to the threat of being vulnerable; it is not synonymous with invulnerability but rather with adaptability and the self-reliance of (local) community. Moreover, adaptability and self-reliance are produced via diversity, modularity and the tightness of feedbacks. A key issue is how normative concepts such as (global) responsibility play out in the spatialisation of Transition Culture resilience: what conceived, perceived and lived spaces might be produced?

The Camp for Climate Action

The first action of the Camp for Climate Action (CCA) in the UK was the 2006 camp at the Drax coal-fired power station in North Yorkshire. The camp was attended by around 500 people who, following several days of communal living, workshops and training, took direct action against Drax, attempting to cause the station to be shut down. As a UK-wide network, CCA have since organised similar camps near Heathrow Airport to oppose plans for a third runway (2007), at Kingsnorth on the Medway estuary in Kent, the proposed site of a new coal-fired power station (2008), and on Blackheath Common, focussing protest on "the city," London's financial district (2009). In 2010 the CCA registered its opposition to the funding of projects with large carbon emissions, overtly targeting the Royal Bank of Scotland, a persistent offender whose major shareholder is the UK government. On August 19th the CCA "took the site on RBS HQ" at Gogarburn in Edinburgh.

Back in 2009, CCA organised a "swoop," a direct action convergence, to close E.ON's Ratcliffe-on-Soar power station in Nottingham. Another swoop the same year saw Campers (literally) pitching-up in the city of London as part of "G20 Meltdown" protests around the meeting of the international economic forum, which targeted financial institutions (2009). In addition, there was a camp at Mainshill Wood in Scotland, the site of a proposed new open-cast coal mine. Also in 2009 Climate Camp Cymru (CCC) organised the first Wales Camp near Ffos-y-Frân opencast coalmine in Merthyr Tydfil, following up with a camp and action at Nant Helen opencast mine in 2010. Along with encouraging regionalisation via local groups and action, CCA also played a major part in the international Climate Justice Action (CJA) network which converged on the COP15 summit in Copenhagen in 2009.

Although some camps and actions are outside the metropolis, targeting sites of high carbon emissions that are both physically significant and highly symbolic, many protest sites are in urban areas and the CCA is comprised of largely urban-based activists. Moreover, the political analysis of CCA identifies capitalism as the root cause of climate change:

> The climate crisis cannot be solved by relying on governments and big businesses with their "techno-fixes" and other market-driven approaches. Their grip on political and economic power lies at the heart of the problem, stifling the development of genuinely sustainable technologies and denying those most severely affected the opportunity to speak up for climate justice. We must therefore take responsibility for averting climate change, taking individual and collective action against its root causes and to develop our own truly sustainable and socially just solutions. We must act together and in solidarity with all affected communities – workers, farmers, indigenous peoples and many others – in Britain and throughout the world (CCA, 2009).

So, one set of targets is the institutions funding and furthering economic growth. The CCA seeks to publicly identify such institutions and to name and shame them. Typically, such institutions are located in the cities (banks, carbon trading companies, pension funds investing in polluting industries, and so on), which, as Massey reminds us, can collectively act as command-and-control centres for the production of socio-environmental vulnerabilities in other places. A rallying call of CCA exemplifies the movement's analysis of the vulnerability faced by communities: "System change, not climate change!"

Commentators variously claim CCA grew out of Earth-First (EF!), the roads protest movement of the 1990s, and the Reclaim the Streets initiative. In particular, the idea of a camp practising and demonstrating sustainable living seems to have been adopted from the eco-village setup in Stirling during the 2005 anti-G8 mobilisation in Scotland (see Harvie et al., 2005). Perhaps the master trope of the CCA is *sustainability*, understood in its fullest sense as heralding environmental, social and economic justice.[2] Certainly, the CCA brings together militant critics of capitalism and environmental destruction, the red and the black along with the green.[3] In the process of this convergence, the CCA has the potential to develop a powerful hybrid resistance, displacing the political boundaries between participant groups through a rigorous process of consensus decision-making that actually centres on encouraging and facilitating dissensus.

---

2   Contrast this with the master trope of "resilience" deployed by Transition Culture, which arguably attempts to side-step justice issues.

3   Red, black and green here representing the range of anarcho-communist, anarchist and environmentalist groups.

The Vulnerabilities of Carrying on Camping

Vulnerabilities are constructed and circulate in the CCA producing various space relations. The vulnerabilities produced by capitalism, principally the vulnerability to climate change of disempowered peoples, is what mobilizes the movement, galvanising it into action. Thence, particular vulnerabilities are constructed in seeking to oppose capitalism. Strategic physical vulnerabilities are deployed in, for instance, non-violent direct action (NVDA). Of course, the movement is also vulnerable to internal fracture and division.

The analysis of the CCA links capitalism inexorably to the extraction of finite fossil fuel resources and climate change. First and foremost, the movement is a response to vulnerabilities constructed by capitalism, specifically, but not exclusively, those attributable to the anthropogenic emissions of carbon dioxide. This vulnerability is the mobilizing ethic of the CCA, a solidaristic rather than altruistic ethic. Arguably, the associated space relation arises predominantly when Campers take direct action which seeks to protect those most vulnerable, or at least afford them a voice. Often, the vulnerable are constructed as the poorest people in countries likely to be hit hardest by climate change, for the most part countries of "the global South" or "Majority World". In Wales, for example, an action featured climate change refugees representing various cultures queuing for re-housing loans from the NatWest bank.[4] Campers also act on behalf of a vulnerable non-human nature, dressing up as polar bears, penguins and even cod to represent the plight of those species.[5] This stewardship is also evident in the CCA's theoretical analysis. Attempting to reformulate the politics of CCA at a national gathering in Bristol in February 2010, one draft statement in particular took on ecological injustice via the goal: "To defend the existence of all species and highlight the critical role of biodiversity." Aware that climate change threatens distant others in time as well as in space, Campers consciously act out of a sense of intergenerational as well as intra-generational injustice, invoking the vulnerability of children, grandchildren, and so on.

Because they set themselves up as against market-driven and government "solutions" to climate change, Campers are immediately vulnerable to the coercive forces of capital and the state. In practice, these forces include both private security companies and the Police. Indeed, this always mistrustful and often confrontational space-relation tends to dominate, distracting campers from others, notably the stated aims of the CCA: *living sustainably and demonstrating alternatives, educating ourselves, taking direct action on climate change, and building a climate justice movement.* Campers' suspicion of government action on climate change extends beyond rejection of market-driven approaches. There is a very real fear of Green authoritarianism and an acute awareness of the vulnerability of always-already contested freedoms.

---

4   NatWest is part of the Royal Bank of Scotland group.
5   An quasi-embodied form of environmental stewardship (see Callicott, 1994).

The vulnerability to the coercive forces of the state and capitalism extends to a cultural vulnerability, whereby Campers are sometimes scorned by an unsympathetic public, often for obstructing their economic behaviours of production, supply, and consumption (see Chatterton, 2006). Elements of the public can indeed be openly hostile. The 2009 Climate Camp Cymru was visited by a group of men who ultimately turned abusive and violent. Aware that such violent relations are a product of a society removed from the practices of the CCA, Campers nevertheless formed a "Tranquility Team" to quell any future disturbances through communicative means.[6] Surprisingly perhaps, the CCA is often most vulnerable to those agencies of the state charged with public health and safety. A camp is as likely to be disbanded for failing to meet health and safety regulations with respect to food preparation, sanitation or risk of fire enforced by bureaucrats, as evicted by Police for trespass. In 2010, Climate Camp Cymru was evicted from a site by a Police operation for driving tent pegs into the ground on the site of a Roman fort, an open-field site used for agricultural production and so routinely worked by heavy machinery.

Campers deploy their vulnerability to coercive forces strategically, using tactics such as "locking-on" to place their bodies at the mercy of capitalism's enforcers. Locking-on is the procedure of making one's body difficult to move, either by attachment to property or fellow activists. To make removal difficult, locks are either strong, for example bicycle D-Locks, typically attaching the activist by the neck to plant or machinery, or concealed as when activists attach themselves to each other by the wrist within a "lock-on tube," which may be made of a matrix of wire and cement and so be resistant to standard cutting tools. Variations of the lock-on tube permit numerous activists to attach themselves to one another. More simply, activists use super-glue to fix themselves to roads, gates and the like. In the city of London swoop in 2009, where newspaper vendor Ian Tomlinson died after being stuck by a Police officer.[7] Campers pitched their flimsy pop-up tents to establish a territory of habitation as protection from overwhelming coercive force.[8] In Merthyr in 2010, activists literally put their bodies on the line to halt trains carrying coal between Ffos-y-Frân mine and Aberthaw power

---

6   In other words, Campers took responsibility for a manifestation of violence that was the product of a political system antithetic to that which the CCA as a movement practices, that is, consensual and non-violent.

7   Following a decision of the Crown Prosecution Service, backed by the Attorney General, the Police officer concerned will not face criminal charges.

8   Though tents are materially flimsy and offer little physical protection, the invisibility they afford can conjure vulnerability in the enforcers of capitalism and the state. At Heathrow, for instance, a large number of protesters remained concealed in a large marquee, breaking out at the opportune moment to out-manoeuvre the Police. Conscious of the transitive power of its mystery, the first task in setting up any Climate Camp is erect a large central marquee, which typically serves as a meeting space.

station,[9] a particularly inefficient plant burning particularly dirty coal. Such tactical deployment of personal, physical vulnerability is a common tactic among peace, justice, and environmental activists engaged in NVDA (see Sharp, 1973). Increasingly, activists are aware of the concomitant psychological vulnerability of engaging in direct action and Activist Trauma Support is set up for precisely that reason:

> (A)s a movement we have not sufficiently acknowledged the psychological effects of the brutality and stress that an increasing number of us are subjected too. Supporting people who have been traumatized should be a central part of our activism (ATS, 2010).

As with all social movements, the CCA is vulnerable to internal strife and division. Prior to the national gathering in Bristol in 2009, to spark discussion, *Shift Magazine* and *Dysophia* published "Criticism without critique – a climate camp reader." The central argument of "Criticism without critique" is that the radical edge of CCA has been blunted by an increasingly "liberal, reformist approach." As an example of the vulnerability to division, one contributor, "a g.r.o.a.t.," argues that the CCA setting out to be as inclusive as possible has in practice meant including middle-class liberals rather than radicals, the working class, the marginalised and disempowered. A g.r.o.a.t. claims there is a tendency for some elements of CCA to ignore neighbourhoods and grassroots struggles in favour of spectacular national action:

> The world does not need another media-savvy Greenpeace; it needs a genuine from-below movement that engages a working class fed up with being patronised and told how much they need to suffer for the benefit of the planet (p. 16).

Mother Earth Rights

While the CCA continues to strive to extend solidarity to vulnerable groups, the working class, the exploited – particularly in the global South, local communities under threat (for example, Sipson Village residents, where some 700 houses, a primary school, historic buildings, and several pubs and shops were threatened by Heathrow Airport expansion, and Residents Against Ffos-Y-Frân in Merthyr exposed to the noise and dust pollution of opencast mining within 37 metre of some urban residences), the network has only begun to extend its politics beyond anthropocentric relational space to encompass a wider nature. Despite the efforts of the CCA gathering in Bristol to formulate a more ecocentric politics, "Criticism without critique" (*Shift*, 2009) bears testimony to this assertion, which arguably holds true for a wider Green anarchism. In response to the farce of

---

9   This action echoed the 2008 "hi-jacking" of a coal train heading for the Drax power station.

the COP15 summit in Copenhagen in 2009, President Evo Morales of Bolivia took the initiative to host the World People's Conference on Climate Change in Cochabamba in April 2010. The Conference produced the *Universal Declaration of the Rights of Mother Earth*, which Morales delivered to the United Nations. Part of Article 1 of the Declaration claims: *Mother Earth and all beings are entitled to all the inherent rights recognized in this Declaration without distinction of any kind, such as may be made between organic and inorganic beings, species, origin, use to human beings, or any other status* (PWCC, 2010). It remains to be seen how the concept of Mother Earth Rights emerging from that conference will play in the generally prosaic and secular CCA, whether the likely dissensus will increase the vulnerability to division and fracture or stimulate a constructive discussion of a more radical green democracy.

The Persistence of Utopian Urbanism

The urbanism circulating in the CCA can be regarded as a form of what David Pinder labels "utopian urbanism." Pinder views the utopian impulse as an irrepressible part of the human spirit, linking contemporary suppression of that impulse with the failure of socialism and also the decline of modernist urbanism. Drawing on the canonical work of David Harvey particularly, Pinder asserts the potential for creative urban thinking to progressively transform cities and processes of urbanisation, rather than accepting these as dystopian spaces either to be ignored in conceptions of the desirable future or studied only in terms of elite practices of escape from and within the city in the here and now (Pinder, 2002, p. 231). Creative urban thinking, Pinder suggests, means asking John Gold's vital question: "What sort of city for what sort of society?" (*ibid*, p. 232). He cites Kevin Robbins' view that the crisis of the city and urbanity is associated with both the scale of physical and social problems, "including the ways in which inequality, segmentation and alienation have been inscribed in contemporary urban landscapes" (*ibid*, p. 232, emphasis added), and also conceptions of the city.

Pinder claims that *a critical* utopian urbanism can counteract prevailing political pessimism and cynicism via its potentially disruptive and transgressive qualities: utopian urbanism need be neither compensatory nor authoritarian. He suggests the potential of developing "modes of critical and transformative urbanism that are open, dynamic and that, far from being compensatory, aim to estrange the taken for granted, to interrupt space and time, and to open up perspectives on what might be" (*ibid*, p. 229). Viewing utopian urbanism not as a unifying vision or a singular emancipatory project, Pinder suggests seeking out the possible in the conditions of the present as means of making multiple interventions in space and time. He claims that "a loss of utopian perspectives in their entirety has disturbing political and cultural consequences, not the least of which is a narrowing of critical thought and a moving away from the *anticipatory moment of critique*" (Pinder, p. 230).

The CCA's utopian urbanism is not a preparation for resource shortages and disaster. Rather, it avoids the millenarian trap and maintains a focus on mitigation of the causes of crises rather than adaptation. However, the CCA does anticipate a heightened vulnerability produced by the likelihood of an increasingly authoritarian state response to these crises. For instance, from an engaged activist perspective, Paul Chatterton provocatively dismisses the environmental crisis to highlight more fundamental crises of justice and democracy, asking: "As good eco-citizens are we sleepwalking into future green prisons, where the EcoRepublic, as the late philosopher Val Plumwood called it, forms a khaki green, quasi-police state using restrictions to save us from climate change? Is the price of having a future, bondage by new green chains?" (Chatterton, 2009).

The CCA is acutely aware of the threat of austerity measures and the vulnerability of those they will likely hit hardest, the poorest and the working class. This has made the movement vulnerable to division as more civic republican minded environmentalists have promoted state-based measures under its banner, measures that are antithetic to CCA's anarchist core. This vulnerability boiled over into emotional conflict during the Heathrow camp (Jasiewicz, 2008; Monbiot, 2008). The CCA perceives climate change as threatening distant others in space and time and so deploys a solidaristic notion of vulnerabilities. The movement asserts its key guiding concept not as austerity, resilience, not even as altruism or sustainability, but as *climate justice*, a subset of environmental justice, promoting the rights of those most affected by climate change, demanding the repayment of ecological debt, and highlighting the critical role of biodiversity. As such, the movement's urbanism tends towards ecoptian urbanism.

Conclusion: Working Towards a Convergent and Progressive Politics of Vulnerability

We start our conclusion with a call that was made following the Camp for Climate Action in 2008:

> So, when you "political" activists come home from this year's climate camp, please seek out local Transition initiatives and bring to them your inspiration and experience. What's in it for you? Here's your chance to construct climate camp every day, to make it the carbon neutral, consensual and creative community in which you live your life! Here are the people who will help you and be helped by you, and together you are stronger. At the same time, let Transition initiatives seek out and welcome home the climate campers. All societies have their warriors, farmers, artisans, artists and groups of every sort. Being open to and welcoming diversity is what makes a society responsive and resilient (Mason, 2008).

It is perhaps significant that the 2008 Climate Camp was convened at Heathrow Airport. This call for a unifying politics of Climate Camps and Transition Towns was made in an attempt to try and find more permanent homes for Climate activists, but also in the hope of attempting to radicalize the Transition Culture movement. The subsequent emergence of *Grow Heathrow* in Sipson that we described at the beginning of this chapter may represent the tentative beginning of this political convergence. It underscores the dynamic transgressive, and emancipatory potential of a dynamic and open living space to produce social life differently. It is in the context of these emerging political circumstances that this chapter has sought to analyze how vulnerability is being mobilized by different urban movements in the UK today, and to reveal how these mobilizations are connected to underlying struggles in and through space.

What clearly unites both the Transition Culture movement and the Camp for Climate Action is that they embody progressive mobilizations of urban vulnerability. That is, they attempt to use vulnerability (defined and realized in different ways) as a basis for calling for greater social and environmental justice within the city. But thinking about these two movements spatially reveals significant differences, and shortfalls, in their operational dynamics. To use Lefebvrian terms, it is clear that the Transition Culture movement utilizes the representational spaces of everyday life as a context for both revealing how people have become vulnerable to the twin threats of peak oil and climate change, and to suggest the power that they hold to build collective forms of resilience to these perils. What Transition Cultures appears to lack, however, is a political vision that can oppose the forces which impose powerful representations of space on urban communities that only serve to perpetuate contemporary forms of socio-environmental vulnerability (this is the crucial connection point between the external and internally constructed modalities of vulnerability we identify). In contrast, it is clear that the transitory nature of Climate Camps means that they lack the emotive, everyday core (what we term experiential vulnerability), which sustains Transition cultures and makes them meaningful for a wide range of people. What Climate Camps do exhibit much more readily is a clear desire to oppose the forces of capitalism and state bureaucracy that create and sustain systemic forms of urban vulnerability both locally and remotely.

Of course, the desire to develop a kind of progressive urban politics that combines the energies, motives and enthusiasm of the everyday life, with a more farsighted resistance to the disheartening structures that shape metropolitan landscapes, has been a long-held goal of the urban left (see Harvey, 1996; Merrifield, 2002). But considering this broader political desire in relation to questions of vulnerability opens up some interesting practical and intellectual insights. It appears that the formation, recognition, and politicisation of vulnerability offers important opportunities for rethinking the nature of the urban condition in the twenty-first century. The use of vulnerability as a progressive political tool does, however, come at a cost: as it spawns new regimes of urban anxiety, neurosis, and potential trauma. It is at this psychological threshold that the paradox of vulnerability politics is expressed. This is a paradox that rests on the need to establish the real and present danger of vulnerability, as a necessary spur for local political action, while avoiding a sense of the inevitable nature of the impeding peril, which could stifle more radical political desires for change. The paradox can surely only be resolved in the practice of new forms of urban politics which seek to connect the vulnerable here with the susceptible there. By shifting the geographical focus of the politics of vulnerability in this way it may be possible to start to think of how urban communities can reshape their own spatial futures, but also change their role in structuring the opportunities that other places have to achieve their own security.

References

ATS (2010) Welcome. Activist trauma support. 2010, 20 August. <http://www.activist-trauma.net/>.

CCA (2009) What unites us? Camp for climate action UK. 2010, 20 August. <http://www.climatecamp.org.uk/>.

Chatterton, P. (2006) Give up activism and change the world in unknown ways: or, learning to walk with others on uncommon ground. *Antipode*, 38, 259–281.

Chatterton, P. (2009) There is no environmental crisis: the crisis is democracy. *Red Pepper.*

Comfort, L.K. Cities at risk: hurricane Katrina and the drowning of New Orleans. *Urban Affairs Review*, 41, 501–516.

Davis, M. (2000) *Ecology of Fear: Los Angeles and the Imagination of Disaster.* London, Picador.

Gibson-Graham, J.K. (2003) Enabling ethical economies: cooperativism and class. *Critical Sociology*, 29, 123–161.

Harvey, D. (1996) *Justice, Nature and Geogrpahy of Difference.* Blackwell, Oxford.

Harvie, D., Milburn, K., Trott, B. and Watts, D. (eds) (2005) *Shut Them Down: The G8, Gleneagles 2005 and the Movement of Movements.* Dissent! and Autonomedia.

Hirsch, R.L., Bezdek, R., and Wendling, R. (2005) *Peaking of World Oil Production: Impacts, Mitigation, and Risk Management.* Department of Energy [US].

Hodson, M. and Marvin, S. (2009) Urban ecological security: a new urban paradigm? *International Journal of Urban and Regional Research*, 33, 193–215.

Hopkins, R. (2008) The rocky road to a real transition: a review. Transition culture. <http://transitionculture.org/2008/05/15/the-rocky-road-to-a-real-transition-by-paul-chatterton-and-alice-cutler-a-review/>.

Hopkins, R. (2008) *The Transition Handbook.* Totnes, Green Books.

Hopkins, R. (2008) Transition towns: a response. *Peace News.* London.

Hopkins, R. and Lipman, P. (2009) The transition network ltd.: who we are and what we do. Transition Network Ltd.

Jasiewicz, E. (2008) Time for a revolution. Comment is free. <http://www.guardian.co.uk/commentisfree/2008/aug/21/climatechange.kingsnorthclimatecam Top of Form 1>.

Kaika, M. (2003) Constructing scarcity and sensationalising water politics: 170 days that shook Athens. *Antipode*, 35(5), 919–954.

Klein, N. (2008) *The Shock Doctrine*. London, Penguin Press.

Lefebvre, H. (1991) *The Production of Space*. Oxford, Blackwell.

Mason, K. (2008) When climate camp comes home. *Peace News.* London.

Mason, K and Whitehead, M (forthcoming) Transition urbanism and the contested politics of ethical place-making. Antipode.

Massey, D. (1992) Politics and space/time. *New Left Review*, 196, 65–84.

Massey, D. (1994) *Space, Place and Gender*. Cambridge, Polity.

Massey, D. (2004) *For Space.* SAGE.

Massey, D. (2007) *World City*. Cambridge, Polity.

Merrifield, A. (2002) Henri Lebvre: a socialist in space. In: M. Crang and N. Thrift N. (eds) *Thinking Space.* Abingdon, Routledge.

Molotch, H. (1976) The city as a growth machine: toward a political economy of place. *The American Journal of Sociology*, 82, 309–332.

Monbiot, G. (2008) Identity politics in climate change hell. <http://www.monbiot.com/archives/2008/08/22/identity-politics-in-climate-change-hell/>.

Newman, P., Beatley, T., and Boyer, H. (2008) *Resilient Cities: Responding to Peak Oil and Climate Change*. London, Island Press.

Pinder, D. (2002) In defence of utopian urbanism: imagining cities after the "end of utopia". *Geografiska Annaler*, 84B, 229–241.

PWCC (2010) Universal declaration of the rights of mother earth. World People's Conference on Climate Change and the Rights of Mother Earth. 2010, 20 August 2010. <http://pwccc.wordpress.com/programa/>.

Seyfang, G. (2009) Green shoots of sustainablity: the 2009 UK transition movement survey. Norwich, University of East Anglia.

Sharp, G. (1973) *The Politics of Nonviolent Action*. Boston, Porter Sargent.

Sieverts, T. (2003) *Cities Without Cities: An Interpretation of the Zwischenstadt.* Spon Press, London.

Shift (2009) Criticism without critique: a climate camp reader. *Shift Magazine & Dysophia.*

Solnit, R. (2009) *A Paradise Built in Hel.* London, Penguin Books.

Trapese Collective (2008) *The Rocky Road to Transition: The Transition Towns Movement and What it Means for Social Change*, Trapese Collective.

Whitehead. M. (2011) The sustainable city an obituary: critical reflections on the future of sustainable urbanism. In: Flint, J. and Raco. M (eds) *The Future of the Sustainable Cities. Critical Reflections*. Policy Press, Bristol 29–46.

Part 3

Vulnerabilities in the Urbanizing Context: Cultural and Demographic Transformations

Chapter 8

Co-opting Restoration: Women, Voluntarism, and Insurgent Performance in Philadelphia

Alec Brownlow[1]

Introduction

Between 1998 and 2003, residents of the predominantly African American neighborhood of Cobbs Creek in west Philadelphia demonstrated a significant and sustained degree of civic environmentalism, volunteering their time, and labor towards the restoration of a local park's natural environment in unprecedented numbers and with a degree of commitment comparable to or exceeding simultaneous efforts occurring around the rest of that city's Fairmount Park System. Local initiative in Cobbs Creek is, at first glance, unremarkable, especially when we consider:

1. the rapid growth of the voluntary sector under advanced liberalism, especially in the area of urban service provision (Wolch 1990; Fyfe and Milligan 2003);
2. the emergence of a new politics of responsibility, citizenship, and community permeating policy and political discourse at all scales (Rose 1999; Raco and Imrie 2000; Taylor 2007); and
3. research suggesting comparable degrees of environmental activism, awareness, and concern among blacks and whites in the US (Jones and Carter 1994; Mohai and Bryant 1998).

Upon closer inspection, however, there is actually little historical evidence to suggest the degree of effort and commitment to nature restoration demonstrated by Cobbs Creek locals; in fact, quite the opposite. First, the residents of Cobbs Creek had, for years, demonstrated little collective activism or organized interest in park matters, generally, much less in questions or issues pertaining to the park's natural landscape or ecological health. Indeed, for the better part of two decades, the park – especially its forested interior – had been altogether abandoned by the community and for many represented little else but a place of anxiety and fear

---

1   The author would like to thank the editors and the participants of the Berkeley Workshop on Environmental Politics for their helpful suggestions on earlier editions of this chapter.

(Brownlow 2006a). Second, the residents of Cobbs Creek had long decried the Fairmount Park Commission as among the more discriminatory and neglectful of Philadelphia's governing institutions, accusing it of systematically underfunding and redirecting resources from parks in black neighborhoods towards those in the city's whiter and wealthier north (cf. Koehler and Wrightson 1987; Wolch et al. 2005; Heynen et al. 2006a). Third, the residents of Cobbs Creek were neither consulted nor invited to participate in the restoration planning process which instead occurred in the distant offices of the Fairmount Park Commission and its partners. Rather, local participation was planned into the restoration process *ex post facto* in the form of unpaid, unskilled, voluntary labor, thereby rendering any claims of local empowerment or collaborative partnership as little more than political rhetoric (cf. Atkinson 1999; McInroy 2000).

And yet, volunteer they did, both in impressive numbers and with a significant degree of commitment and initiative. This chapter begins to explain this apparent contradiction. I approach voluntarism as a historically informed, spatially contingent, and embodied political performance, one aimed both at drawing attention to and redressing a repertoire of identity-specific injustices that have effectively eliminated park access among many local Cobbs Creek residents. In particular, my focus in this chapter is the participation of women in the restoration process. Informed by feminist analyses of women's spatial (re-) appropriation and politics of place in the global city (e.g., Lind 1997; Wekerle 1999; Harcourt and Escobar 2005), I argue that, for the women of Cobbs Creek, the motivations and incentives to (re-) enter the park and volunteer their time and labor towards its ecological overhaul are political and intensely personal (see Gibson-Graham 2005), derived from intersecting experiences with (and vulnerabilities to) state led racialized neglect and escalating rates of male sexual violence locally. In Cobbs Creek, the intersection of racism, sexism, and their respective violences and exclusions is perhaps nowhere more apparent or more thoroughly represented than in the local park's landscape. Among local women, their self-imposed exile from Cobbs Creek Park, and the vulnerabilities and injustices framing this absence, constitute the context underpinning their participation in its restoration. I argue, first, that restoration presents local women with an otherwise rare, if ultimately ephemeral, opportunity for secure re-entry into a public space of historical significance in the everyday politics and practices of social reproduction; moreover, restoration offers to local women the opportunity to participate in the landscape's material overhaul, all the while working within and through state-sanctioned policies and discourses. In the process, park-based voluntarism opens up political space from which local women can insurgently claim their rights to the city and to citizenship by making visible and challenging the conditions of their exclusion (Bondi 1998; Ruddick 1996). I argue that their voluntary participation in Cobbs Creek Park's restoration may thus be considered a kind of place-based identity politics, one that calls attention to and ultimately challenges racialized histories of injustice and neglect and the subsequent exclusions and vulnerabilities that accompany them.

Second, I suggest that the act of restoration itself presents to local women the opportunity to, individually and collectively, respond to, invalidate, and perhaps reverse environmental changes marked as patriarchal and exclusionary and to re-create a local landscape that is inclusive of and accessible to women, children, and the everyday rights and relationships of social reproduction. To this end, my argument both draws from and follows critical and libratory traditions in feminist geography that reveal and challenge women's marginalization from, vulnerability within, and resistance to geographies marked by patriarchy, racism, and class (e.g., Bondi and Domosh 1998; Carney 2004; Fincher et al. 2002; Gibson-Graham 2006; Harcourt and Escobar 2005; Rocheleau et al. 1996; Ruddick 1996; Schroeder 1999). Consequently, following the historical development of "traditional" (i.e., "third world") political ecology before it (Robbins 2004), the intent of this chapter is to introduce feminist analysis and theory to a still-emerging urban political ecology (Swyngedouw and Heynen 2003; Heynen et al. 2006b).

The chapter has two principle aims. First, I hope to expand upon current debates on the political meaning and implications of voluntarism and its widespread growth as an instrument of urban service provision at the turn of the twenty-first century, and to suggest that dominant representations of voluntarism as either "empowered participation" (Fagotto and Fung 2006) or as the latest instrument state regulation and hegemony (Kearns 1992; Raco and Imrie 2000) are incomplete insofar as they: 1) respectively embrace and critique voluntarism using the term's most normative interpretation as it has been shaped within the larger narrative of neoliberalism; and 2) fail to explore the possibilities for co-optation by groups with alternative political goals and agendas (Jones 2003; Mayer 1995). Alternatively, I adopt Holston's (1999) concept of "insurgent citizenship" and social theories of political performance (Houston and Pulido 2002; Gilbert and Phillips 2003) as especially powerful and *apropos* in their combined ability to account for and identify those practices and spaces available to subaltern groups to disturb hegemonic and exclusive representations of society and disrupt normative qualifications of citizenship in their respective efforts to (re-) claim political rights and (re-) appropriate political space (see Young 1990).[2] The lens of insurgent performance moves beyond simplistic and unsatisfying descriptions of "eco-stewards" (Jordan 1994; Miles et al. 1998; Schroeder 2000) or "eco-citizens" (Light 2003) to effectively bring the analysis of women's voluntarism in Cobbs Creek into the political arena of rights and resistance. It also allows for a more contingent and robust political agency than that generally identified with the more normative understanding of the "active citizen" (see later; Kearns 1992). Rather, as I will argue, the embodied place-based performance of landscape restoration creates the political space necessary to expose, address, and ultimately disrupt the historical and often invisible injustices of racialized and gendered vulnerabilities. Individually and combined, insurgency and performance are fitting

---

2 Using this definition, Holston's "insurgent space" or "spaces of insurgency" (1990:37) is very much akin to Lefebvre's (1990) "representational space".

tropes to describe women's co-optation of restoration and its voluntary activity as a gendered and race-based means of political expression and activism (Wekerle 1999).

Second, I hope to expand upon recent claims of discursive Nature's re-emergence and re-signification in urban policy and planning as a means for capital accumulation, ecological gentrification, and/or social regulation (Keil and Graham 1999; Whitehead 2003; While et al. 2004; Dooling 2009). While these are powerful and undoubtedly legitimate arguments indicating the ever-widening narratives and ever-encompassing geographies of urban revitalization and competitiveness, there is as yet little-to-no attempt to conceptualize or identify how the top-down discourse of Nature may be subverted and co-opted by subaltern groups in their efforts to achieve local goals of political inclusion and access, rather than simply a narrative to be struggled against (Desfor and Keil 2004). This kind of resistance to the production and maintenance of vulnerabilities is perhaps nowhere more apparent than in urban parks, whose hybrid identities as urban nature and public, political space make them especially prone to contestation and co-optation. Insofar as it continues to depend on a volunteer labor force, restoration, as I show in this paper, offers just such an opportunity for subaltern political activity. Again, I turn to insurgency and performance as especially relevant and meaningful expressions of political agency in urban landscapes whose discursive identities and governance structures are increasingly mobilized to meet the economic conditions and demands of globalization and entrepreneurial urban restructuring (Harvey 1989; Hubbard and Hall 1996; Desfor and Keil 2004).

The chapter begins with brief introductions to voluntarism and participation, as these notions are interpreted and conceptualized within the larger narrative of neoliberalism and state decline. I discuss the various debates over voluntarism's (and, by extension, civil society's) meaning and function, as either an instrument of local empowerment or of state authority, and introduce the concepts of insurgency and performance as each assists in clarifying voluntary activity as a political, historically, and geographically contingent activity. The next section introduces the west Philadelphia case study and analyzes and discusses voluntary activity in Cobbs Creek Park using these concepts.

Voluntarism: Performance and Insurgency

Urban economic restructuring and the institutionalization of neoliberal economic and political reforms have ushered in a new era of civil society; social services whose provision and distribution were once the domain of the state are increasingly re-framed as outside the responsibility of government and re-located, or devolved, to a rapidly growing sector of volunteer organizations, non-profits, and religious institutions. Today, the turn to civil society and local governance (vs. government) as necessary institutional fixes in response to "the crisis of Fordism" is pervasive, reflected in the now normative narratives of voluntarism, participation, and local knowledge that currently permeate western political discourse. As Wolch (1990) suggests, the growth of the voluntary sector, in particular, has been embraced both by the political left and the political right. For the right, the widespread move to voluntarism in western society represents a long overdue move away from the welfare model of an inefficient and bureaucratic government apparatus. It represents on both sides of the Atlantic the ideological and discursive success of small government and local responsibility ushered in during the Reagan and Thatcher administrations of the 1980s, honed and made more socially appealing during the Clinton and Blair administrations of the 1990s (Peck and Tickell 1992; Tickell and Peck 1995; Raco and Imrie 2000). Alternatively, among the left, the move to civil society and voluntarism represents opportunities for local capacity building and empowerment; an almost utopian model of deliberative democracy structured around community-based governance and decision-making that are informed by and representative of local knowledge and experience (e.g., Fagotto and Fung 2006).

Significantly, left and right interpretations of voluntarism and civil society are not mutually exclusive. Their general overlap and broad approval in political discourse has expedited their establishment as normative instruments of local governance and, by extension, citizenship, whereby the "good" (or *active*) citizen picks up where the state no longer functions and donates his or her time, labor, skills, and knowledge to the betterment of a just and democratic society. In short, the discourse of civil society and its affiliates (citizenship, voluntarism and participation) has become political hegemony (Hay 1995; Goodwin and Painter 1996). This has raised flags among many social theorists who doubt seriously the unproblematic, benign release of state power and authority to fragmented, diverse, and perhaps oppositional local interests. Rather, to use Wolch's (1990) expression, the voluntary sector takes on the role of a *shadow state*, whereby the state – through a variety of institutions and devices – retains discursive control over rhetorically independent local voluntary activities as a means of controlling any potential for political opposition and ensuring its own reproduction. This can be assured through intrusive devices like public–private partnerships, state control of competitive state funding and resources, etc. Political economic assessments of civil society have, in particular, adopted this angle of critique (Harvey 1996; Atkinson 1999). A more radical assessment is offered by Raco and Imrie (2000), whose Foucauldian analysis posits that the successful transition in political discourse to civil society has involved the widespread creation of new subjectivities of self-governance and self-regulation. From this perspective, *active* citizenship and the "good" citizen are discursively linked to the concepts of (local, political) "responsibility" and "accountability" whose normative meanings and interpretations are easily manipulated by and not ideologically dissimilar to those of the state. Here, the devolution of responsibility by the state to the local scale, discursively reframed by the state as "active citizenship", "local governance" and "deliberative democracy" creates the conditions for its own reproduction (cf. Goodwin and Painter 1995; Jessop 1995; Tickell and Peck 1995).

There remains, however, the still powerful and popular argument from the left that the ceding of decision-making and action by the state to the local level presents opportunities for local empowerment and capacity building (cf. Young 1990). While political–economic and poststructural critiques have successfully cast suspicion upon such claims, there is growing skepticism about their dismissal or neglect of political agency and the possibility of political co-optation. For instance, Jones (2003) and Mayer (1995) – while raising their own doubts on civil society's supposed independence – are nonetheless critical of growing calls among the radical left to abandon participatory discourse insofar as it is, may be, or has the potential to be corrupted by state influence (e.g., Rose and Miller 1992; Raco and Imrie 2000). For both, this amounts to throwing the baby out with the bathwater. Rather, they take a more pragmatic approach, advocating the significance of participation for a subaltern oppositional politics. For them, sweeping claims of the state's regulation of civil society should be considered along with, instead of to the neglect of, the vulnerability of civil society to co-optation and the agency of "active citizens" to challenge advanced liberalism from within.

From this, I now turn briefly to Holston's (1999) notion of insurgency and "insurgent citizenship" as an appropriate expression of subaltern co-optation and struggle. For Holston, the "spaces of insurgent citizenship" are defined as those identities, opportunities, and activities from which subaltern groups can disrupt and challenge normative, ideological, and exclusionary accounts of society and citizenship (Young 1990). Holston (1999) intended his concept of insurgency for an increasingly diverse, fractured, politically and economically marginalized urban immigrant population. As Wekerle (1999) demonstrates in her analysis of gender planning in Toronto, however, "insurgent citizenship" is an unrestricted and adaptable concept, one that can be applied to other groups on the social and political margins, especially women. She uses the idea of insurgency to describe how and where the counter claims and oppositional politics of women in Toronto are being inserted into otherwise patriarchal planning institutions and policies. Indeed, insurgency distinguishes feminist activism against the injustices of patriarchy and economic restructuring in cities globally (Rabrenovic 1995; Lind 1997).

Insurgency is closely aligned to the notion of performance. Houston and Pulido (2002:403) define performance as a form of "embodied revolutionary praxis", one that is "enacted in specific historical and geographical contexts … to expose the dynamics of power and exploitation". Performance, they continue, "operates simultaneously as a space of possibility and becoming, and as a mechanism for working through existing social contradictions [and injustices] by making them visible" (406). Here, the identity of the performer combines with the place of performance towards the production of a robust politics of place. As I argue below, black women's performance *qua* voluntarism in Cobbs Creek Park creates political space from which political claims for the rights of citizenship (including the right to the city; Mitchell 2003) are made, vulnerabilities challenged, and injustices exposed; it also makes visible black women's decades-long absence from the park.

Voluntarism in Cobbs Creek Park is political performance insofar as it reinscribes public space with black women in it. Performance not only facilitates political expression, but is itself a form of identity politics. It is, according to Gilbert and Phillips (2003), an insurgent means to citizenship among the politically marginal – or, what the authors refer to as "performative citizenship". As I argue in this paper, performance is not inherently spectacular (e.g., marches, protests, and demonstrations); it can infiltrate more mundane, everyday practices and activities – like volunteering (see Hobson 2006). Used in this way, voluntarism can be co-opted as an everyday mechanism of resistance to the production of urban vulnerabilities (Scott 1990; see Houston and Pulido 2002).

Cobbs Creek

Cobbs Creek is the largest neighborhood in west Philadelphia (Figure 8.1). Its current history dates to the middle of the last century, when working and middle-class whites fled upon the arrival into west Philadelphia of blacks escaping the ghettos in the city's north and south. A majority white neighborhood in 1950, Cobbs Creek was almost entirely black by 1970. By then, the conditions of segregation and economic and political neglect were gaining a grip upon what, initially, had been a prosperous and politically powerful middle and upper middle class neighborhood. Today, Cobbs Creek is dominated by social indicators of political neglect and economic impoverishment, including high rates of unemployment (especially among working age males), abandoned housing, and families living in poverty. Violent crime is on the rise. Between 1995 and 2003, rape in Cobbs Creek grew by an annual rate of 11 percent, a pace 1.5 times that of the rest of the City. During this same period, murder in Cobbs Creek grew at an annual rate of over 4 percent while the rest of Philadelphia experienced an average annual decline of 6.4 percent.[3] The local Cobbs Creek Park has been gripped by violence for nearly three decades. Over two-dozen men and women have been murdered or discovered dead in the park since 1980, including an especially horrifying period in the 1990s involving mutilation and burning. Rape and sexual violence are not uncommon. The subsequent abandonment of the park by many locals, especially women and children, was pervasive by the late 1980s.[4]

---

3   Crime statistics provided by the Philadelphia Police Department and FBI Uniform Crime Report.

4   Feminist scholarship consistently and correctly decries the repeated misrepresentation of public space and stranger attack as the site of women's risk (Valentine 1989; Pain 1997, 2001). I take this critique seriously and have commented on it elsewhere (Brownlow 2005, 2006a). My focus on public space is in no way intended to reproduce the misrepresentation of sexual violence as primarily a crime of the public sphere.

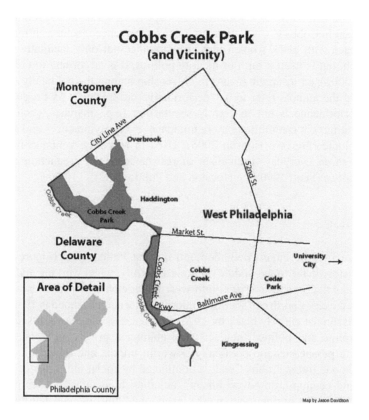

Figure 8.1   Cobbs Creek and Cobbs Creek Park

Cobbs Creek Park is an 800-acre stretch of urban forest following the contours of the neighborhood's (and the city's) western boundary. It is one of seven watershed parks that comprise Philadelphia's Fairmount Park System (Figure 8.2). As the neighborhood's largest and most widely recognized public space, Cobbs Creek Park played a critical role in the growing black community's early social and political development, providing space and opportunity for *ad hoc* leisure and social interaction among generally middle and upper-middle class strangers arriving in the west from the city's northern and southern ghettos. Among women, children, and the elderly, in particular, Cobbs Creek Park was a place for community development and exchange and social reproduction; further, it represented a safe place to avoid the growing tensions and conflicts on "the street" in the 1960s and early 1970s. To many, Cobbs Creek Park was a resource and a haven whose likes had been unavailable in the ghettos of Philadelphia's north and south. The park was also the site of local and regional black political activism and activity, both progressive and radical; from peaceful civil rights demonstrations in the 1960s, to more radical and often less benign gatherings and activities by the local chapter

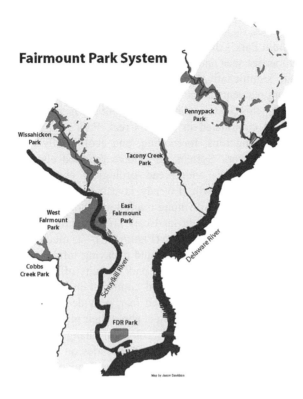

Fairmount Park System

Figure 8.2   Fairmount Park system

of the Black Panthers in the 1970s and MOVE in the 1980s, the park occupies a symbolic position in the community's and the city's black political history.[5]

The politics of park decline and environmental change in Cobbs Creek is complex, a product of social relations and urban economic change that gripped all of Philadelphia and the urban industrial north during the second half of the 20th century (Brownlow 2006a). Nonetheless, among the residents of Cobbs Creek, responsibility for park decline – like neighborhood decline, in general – is placed squarely at the steps of the city government and its affiliated institutions, especially – in this case – the Fairmount Park Commission. Among Cobbs Creek residents there is widespread consensus that their park, like others in the city's

---

5   MOVE was a Philadelphia-based black liberation group housed and headquartered in the Cobbs Creek neighborhood. In 1985, the MOVE home and headquarters was destroyed when an incendiary bomb was dropped by a Philadelphia Police helicopter. Dozens of surrounding homes were destroyed in the ensuing fire and hundreds of people displaced.

black neighborhoods, has suffered from decades of unjust, racially motivated practices and policies, *de jure* and *de facto*, of neglect and resource redistribution, whereby Fairmount Park's dwindling resources are parlayed into those parks and park-owned properties seen and used by whiter and wealthier communities in the city's north and, significantly, by the city's growing population of tourists and young, gentrifying residents in the bustling Center City area (Brownlow 2006a).[6] By the mid-1990s, this 30-year dialectic of political neglect, park decline, and local abandonment culminated in Cobbs Creek Park's reputation as the most misused, polluted, dangerous, foreboding, and ecologically bereft park in the 8900 acre Fairmount Park System. As I have reported elsewhere (Brownlow 2006a,b), in Cobbs Creek, widespread ecological change – especially the unmanaged, uncontrolled growth of weedy, invasive species – is a daily reminder of discriminatory park politics. Among many local women, the interpretation of and response to these changes in the landscape are especially pronounced; to them, the uncontrolled growth of, e.g., kudzu (Figure 8.3) – as much as abandoned cars, trash piles, graffiti, drug paraphernalia, and decaying infrastructure – represents not just the desertion of Cobbs Creek by park and city officials but, significantly, a complete absence of park-based social control. For women, local ecological change amplified feelings of park-related fear, provided further justification for self-exile, and thus reproduced resentment about their marginal position. In short, to the women of Cobbs Creek, local ecological change is patriarchal through and through, representing, on the one hand, decades on the periphery of white Philadelphia's political and economic consciousness and, on the other, a pervasive and exclusionary fear of male sexual violence locally, both for themselves and their children.

It was into this context that restoration waded in the mid-1990s and through which women's voluntarism in the restoration project must be understood.

Restoration

Park relief arrived in 1996 when, in commemoration of its centennial anniversary, the philanthropic William Penn Foundation (WPF) granted the Fairmount Park Commission (FPC) $26.6 million for the express purposes of ecological restoration and environmental education throughout the systems watersheds (see Goldenberg 1999). It was, and remains, the largest award of its kind ever granted for urban park renewal. Ostensibly a project intended to benefit urban ecological systems and enhance the appreciation of urban nature, the significance of park restoration to Philadelphia's struggling economy quickly became apparent, at times overshadowing any ecological imperatives it may originally have entertained.

---

6   Center City is the only neighborhood in Philadelphia experiencing population growth.

Figure 8.3   Ecology of fear: kudzu in Cobbs Creek

The WPF is Philadelphia's and the Delaware Valley Region's most ardent philanthropic booster, contributing millions of dollars annually to projects and organizations involved in efforts of economic revitalization and urban competitiveness. To this end, the WPF is an instrumental institution to the politics and processes of place-making in Philadelphia, funding and investing in projects aimed at attracting inward investment, tourist dollars, and full-time residents back into the city.[7] It is, then, unsurprising that the WPF should turn their attention to Philadelphia's Fairmount Park, one of the oldest park systems in the US urban history and among the largest municipal park systems in the world. Indeed, the growing demand for increased quality of life by a discriminating and increasingly mobile public is arguably the driving force behind the current urban park "renaissance" in North America and Europe (see Harding 1999; Harnik 2000; Thompson 2002; Pincetl and Gearin 2005; also Keil and Graham 1998).

Following the award to the FPC, there was assembled a public–private partnership to initiate, implement, organize, oversee, and administer restoration throughout the park's 8900 acres of greenspace. The three-volume Restoration Master Plan was developed by the research arm of the Philadelphia Academy of Natural Sciences; the project was administered by the newly established Natural

7   Since 1960, Philadelphia's population has declined by nearly 25 percent. The loss of population to the surrounding suburbs has been disproportionately white. According to the 2000 census, whites were in the city's minority for the first time in the city's history.

Lands Restoration and Environmental Education Program (NLREEP), a quasi-independent agency funded by the Penn grant, attached to, housed within, and beholden to the Fairmount Park Commission. As NLREEP's Director, the WPF lured away from her powerful position as Program Director at the Center City District (Philadelphia's largest, most prosperous, and most powerful business improvement district) one the city's most passionate boosters and creative visionaries. The labor intensive task of restoration, involving primarily the removal of non-native, invasive plant species and their replacement with native vegetation, would be the job of a volunteer labor force derived from park-specific communities, community organizations, churches, schools, and, where they existed, *in situ* park-related advocacy groups.[8] Volunteer activities, solicitation, and organization were administered by full-time NLREEP Volunteer Coordinators. So as to avoid the kind of resentment and claims of discrimination from the city's black community that plagued the FPC, it was decided by NLREEP early on that Cobbs Creek would be the first community approached and that Cobbs Creek Park would be the first restored.

Voluntarism in Cobbs Creek

This section explores the dynamics and demographics of voluntarism in Cobbs Creek Park's restoration using data gathered by NLREEP between 1998 and 2003. For purposes of demonstration, data from Cobbs Creek is presented with comparable data from the park system's three principle remaining watersheds: Wissahickon (in the city's northwest); Pennypack (in the city's northeast); and Tacony (in the city's north-central) (Figure 8.2). Combined, these four parks constitute almost half of the Fairmount Park System's land holdings. The parks vary considerably in size, with Wissahickon and Pennypack Parks each larger than Cobbs Creek and Tacony combined. Because of their situation in the city's landscape and their place in the city's history, the parks are also distinguished by different constituencies and local politics. Wissahickon Park (or, the Wissahickon) is the oldest protected watershed in Philadelphia and is considered by many the flagship of the entire Fairmount Park System. Adjacent to it are some of the wealthiest and most powerful neighborhoods in the city (Chestnut Hill, East Falls, and Mount Airy) and, for over a century, it has enjoyed the advocacy and resources of the influential Chestnut Hill-based *Friends of the Wissahickon*. By contrast, Tacony is surrounded by a rapidly changing, increasingly ethnically and racially diverse working and middle-class population in the city's north, while Pennypack's constituency is almost entirely white working and middle-class.

---

8   There are currently over 100 park-related "Friends" organizations, most devoted to the protection and stewardship of a certain portion of the Fairmount Park System or to the protection of certain rights within the park, e.g., bicycling and horse-back riding.

These local, place-based constituencies overwhelmingly constituted the ranks of restoration volunteers in their respective parks.

The degree of, or commitment to, voluntarism is demonstrated in Table 8.1. I define "commitment" as those indicators of initiative emerging from within the volunteer labor pool itself (e.g., number of volunteers per event, volunteer hours, etc.) rather than those outside of local control (e.g., number of restoration events). As Table 8.1 indicates, relative to elsewhere in the Fairmount Park System, the volunteers in Cobbs Creek demonstrated considerable commitment to local restoration. Despite having access to a fraction of the voluntary opportunities made available to the Wissahickon constituency (a statistic perhaps explained by differences in park size, as identified in the comparable number of events per park acre), the residents of Cobbs Creek demonstrated considerably greater participation in the opportunities that were made available to them. For instance, restoration events in Cobbs Creek attracted more volunteers on average (almost 23) than any other park (although Tacony runs a close second at 21). For any given event Cobbs Creek volunteers dedicated, on average, nearly twice as much time *per capita* to restoration (3.5 hours) than did the Wissahickon volunteers (1.9 hours). Similarly, the number of volunteer hours (a collective rather than *per capita* statistic) dedicated to each restoration event was higher by a considerable margin in Cobbs Creek (79.6) than anywhere else in the Fairmount Park System. Cobbs Creek also led all parks in volunteer hours per park acre (47) and was second only to Tacony in the number of volunteers per park acre. In short, although these statistics do not indicate the quality of work performed, they nonetheless suggest a sustained and – relative to elsewhere in the Fairmount Park System – notable commitment among Cobbs Creek's volunteers.

Demographically, women were poorly represented among the ranks of volunteers in Cobbs Creek relative to Tacony, Pennypack, and the Wissahickon, constituting less than one-quarter (23.8 percent) of the local volunteer population. By contrast, women's participation was nearly twice as great throughout the rest of the Fairmount Park System, averaging 41.5 percent in the remaining three watersheds, with a high in Tacony of nearly 47 percent (Figure 8.4). This statistic can be partially explained by: 1) the nearly 30-year alienation of local women from and their continued fear of Cobbs Creek Park, 2) ambivalence towards Cobbs Creek Park by the local community, especially among the younger population (i.e.,

Table 8.1   Fairmount Park restoration volunteer statistics (1998–2003)

	Acres	Events	Vols	Vol Hrs	Hr/ Vol	Vol/ Evt	VHr/ Evt	Evt/ Acre	Vol/ Acre	VHr/ Acre
Cobbs Creek	787	466	10656	37113	3.48	22.87	79.64	0.6	13.54	47.16
Wissahickon	1694	1032	16184	30440	1.88	15.68	29.50	0.6	9.55	17.97
Pennypack	1587	284	5643	14504	2.57	19.87	51.07	0.2	3.56	9.14
Tacony	255	199	4336	10581	2.44	21.79	53.17	0.8	17.00	41.49

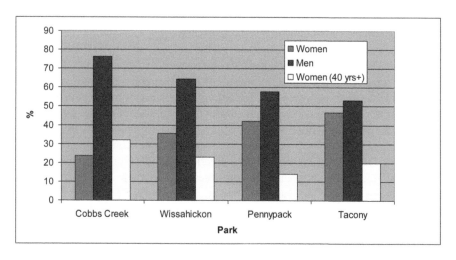

Figure 8.4   Voluntarism and gender in Fairmount Park (1998–2003)

30 years and under), 3) relatively under-developed and politically weak institutions of park-based support and advocacy,[9] and 4) local hostility and distrust towards the FPC, especially among women. However, among those women who did participate in the park's restoration, other interesting patterns appear. Foremost among these is the representation of women 40 years and older who constituted nearly one-third (32 percent) of all female volunteers in Cobbs Creek between 1998 and 2003. By contrast, in the remaining three watersheds, this age group represented, on average, only 19 percent. Several were over the age of 70. The average age of female volunteers in Cobbs Creek was 32 years. In the Wissahickon, Pennypack, and Tacony the average female volunteer was 23.5 years old. Further, a handful of older women were *de facto* leaders in the Cobbs Creek project, marshalling local support and enthusiasm for restoration, summoning and assembling volunteers for NLREEP-organized restoration activities in a way and to a degree the NLREEP staff were unable to. Of note, these women, all in their sixties or seventies, were among Cobbs Creek's original African American pioneers and, on average, had lived in the neighborhood for 37 years. Individually and collectively, they were a fount of local knowledge and experience. They, like few others, knew the history and potential of the park as a viable and vibrant public, political space: one that was critical to the community's embryonic formation and identity in the 1950s and 1960s; one with profound symbolic significance to the politics of racial protest in the 1960s and 1970s; one of historical significance to women's public sphere participation; and, of course, one whose decline and subsequent estrangement

---

9   By contrast, Wissahickon, Tacony, and Pennypack Parks each enjoy the support and advocacy of organized Friends groups.

from the local community has had profound social impacts, especially for women and children.

These women were, I suggest, the heart of voluntarism in Cobbs Creek. Their résumés suggest their political motivations in the park's restoration and in community participation. For example, one is the daughter of a late leader of the local NAACP chapter and was married to one of Philadelphia's most prominent black political activists of the civil rights era. She is the founder of Cobbs Creek's first and only environmental center, an organization whose primary mission – while ostensibly environmental (and thus eligible for a certain amount of financial support from the FPC) – is the reintroduction of local youth to Cobbs Creek Park. Another is the mother of a child murdered along the park's periphery upon whose death she reclaimed an in-park building that had been abandoned by the city, developing it into a recreation and cultural center for local youth. A third, now deceased, single-handedly brought ice-skating to the park in the early 1990s in what is consensually agreed upon as the first attempt to reclaim the park for local use and local youth.[10]

When considered in combination with the relatively impressive voluntary turnout by adult women – many of whom were children during the park's heyday in the 1950s and 1960s – the picture that appears is of first, an insurgent women's "movement" (however unorganized) whose intent is the (re)appropriation of public space via performative (re)occupation (cf. Wekerle 2005) and second, a mother's "movement" aimed at wresting the park away from criminal activity and reintroducing it to local youth whose experience with it is, at best, minimal and fearful, among boys and girls alike (Brownlow 2005; see Rabrenovic 1995). Both represent, I argue, an attempt by local women to reclaim political space from its current patriarchal form. Through their voluntary labor and in-place presence, local black women expose and challenge the park's patriarchal hegemony – as a landscape marked by white political neglect and male sexual violence – and create a feminized politics of place that goes beyond inclusiveness to ascertain rights to the city and to citizenship for local women and local youth, alike. Among land management practices, restoration – despite its normative ecological ambitions and ideological vulnerabilities (i.e., as an exclusionary discourse) – is unique in its permissiveness of (or, at least, its vulnerability to) insurgent subaltern co-optation. Through its philosophical and pragmatic embrace of voluntarism, restoration opens itself up to groups on the political margins as an opportunity for the politics of identity and place. Restoration drew local black women, literally, back into a political space from which they had long been alienated and from which they then could make claims of ownership and rights of access, for both themselves, their children, and their community; voluntarism provided them the performative vehicle from which these claims could be publicized and made legitimate.

---

10   Guffawed by city politicians and park commissioners as an activity that would attract few blacks, the Cobbs Creek ice skating rink has produced several junior Olympians and many city, regional, and state ice skating champions.

Conclusion

Women's social movements, Wekerle (1999, 2005) notes, are at the political fore in challenging the injustices and vulnerabilities of everyday life that have accompanied the growing hegemony of neoliberalism and urban economic restructuring in North America (see also Rabrenovic 1995). Lind (1997) argues similarly about women's social movements in developing world cities. Yet their focus on the everyday has resulted in the relative obscurity of women's social movements in urban social theoretical scholarship. As Wekerle suggests, there is little concern for the decline of everyday life and, by implication, women's political space within current metanarratives of neoliberal urbanism and globalization (2005:87; Harcourt and Escobar 2005). To the extent that this, in fact, is the case, and insofar as the "everyday" is a complex and dynamic fabric of identities, spaces, and social relations, women's urban social movements remain poorly understood and infinitely diverse in both form and structure. Globalization suggests that this complexity and diversity will only continue to grow. In this chapter, I have adopted these premises to help account for civic environmentalism among Philadelphia's most marginalized constituency. I have argued that, on the one hand, voluntarism in the public sphere provides local women opportunities for insurgent and performative claims to citizenship, and to the rights to the city and to nature both for themselves and for local youth (see Gilbert and Phillips 2003). It is, so to speak, an effort to "take back the park", similar in aims, if not in structure, to the performative anti-violence "take back the night" movements seen on city streets. On the other hand, I have suggested that, among urban land management options, restoration is socially and politically distinct, offering to local women the opportunity to repair or reverse landscape level changes marked as patriarchal and exclusionary. Voluntarism, like its affiliates (e.g., participation, civil society, and citizenship) is vulnerable to political co-optation, both from above and from below. While the former – co-optation from above, or by the state – has received generous (and legitimate) scholarly attention and is a significant component of a theoretically sound, robust critique of neoliberal state authority, the latter, co-optation from below, has remained relatively obscure. That this should continue to be the case, especially in global cities whose populations and politics are increasingly fragmented and diverse, is surprising to say the least. But it also offers ample, almost infinite, opportunity – suggested by Jones (2003) and Mayer (1995) – for scholarship and discovery, especially as the voluntary sector's reach continues to broaden, diversify, and move ever deeper into the city's social fabric. As Wolch's (1990) shadow state becomes ever more complex, so do the opportunities for its infiltration and co-optation. Finally, insofar as the voluntary sector's historical feminization (i.e., a patriarchal construction of voluntarism as the normative domain of women) retains any material form or is at all a viable and apt description of the sector today, voluntarism then appears – perhaps more so than ever before – to offer an increasingly diverse population of urban women an especially powerful and spatially pervasive means of political expression, opposition, and struggle for the rights of citizenship.

References

Atkinson, R. 1999. Discourses of partnership and empowerment in contemporary British urban regeneration. *Urban Studies* 36:59–72.

Bondi, L. 1998. Gender, class, and urban space: public and private space in contemporary urban landscapes. *Urban Geography* 19:160–185.

Brenner, N. and Theodore, N. (eds) 2002. *Spaces of Neoliberalism: Urban Restructuring in North America and Europe*. Blackwell, Oxford.

Brownlow, A. 2005. A geography of men's fear. *Geoforum* 36:581–592.

Brownlow, A. 2006a. An archaeology of fear and environmental change in Philadelphia. *Geoforum* 37:227–245.

Brownlow, A. 2006b. Inherited fragmentations and narratives of environmental control in entrepreneurial Philadelphia. In: N. Heynen, M. Kaika, and E. Swyngedouw (eds) *In the Nature of Cities*, pp. 208–225. Routledge, New York.

Carney, J. 2004. Gender conflict in Gambia wetlands. In: R. Peet and M. Watts (eds) *Liberation Ecologies*, pp. 289–308. Routledge, New York.

Desfor, G. and Keil, R. 2004. *Nature and the City: Making Environmental Policy in Toronto and Los Angeles*. University of Arizona Press, Tucson.

Dooling, S. 2009. Ecological gentrification: a research agenda exploring justice in the city. *International Journal of Urban and Regional Research* 33:621–639.

Fagotto, E. and Fung, A. 2006. Empowered participation in urban governance: the Minneapolis neighborhood revitalization program. *International Journal of Urban and Regional Research* 30:638–655.

Fincher, R., Jacobs, J.M., and Anderson, K. 2002. Rescripting cities with difference. In: J. Eade and C. Mele (eds) *Understanding the City: Contemporary and Future Perspectives*, pp. 27–48. Blackwell, Oxford.

Fyfe, N.R. and Milligan, C. 2003a. Out of the shadows: exploring the contemporary geographies of voluntarism. *Progress in Human Geography* 27:397–413.

Gibson-Graham, J.K. 2005. Building community economies: women and the politics of place. In: W. Harcourt and A. Escobar (eds) *Women and the Politics of Place*, pp. 130–157. Kumarian Press, Bloomfield.

Gibson-Graham, J.K. 2006. *The End of Capitalism (as we knew it)*. University of Minnesota Press, Minneapolis.

Gilbert, L. and Phillips, C. 2003. Practices of urban environmental citizenship: rights to the city and rights to nature in Toronto. *Citizenship Studies* 7:313–330.

Goldenberg, N. 1999. Philadelphia launches major restoration initiative in park system. *Ecological Restoration* 17:8–14.

Goodwin, M. and Painter, J. 1995. Local governance, the crises of fordism and the changing geographies of regulation. *Transactions of the Institute of British Geographers* 21:635–648.

Hall, T. and Hubbard, P. 1996. The entrepreneurial city: new urban politics, new urban geographies? *Progress in Human Geography* 20:153–174.

Harding, S. 1999. Towards a renaissance in urban parks. *Cultural Trends* 35:3–20.

Harnik, P. 2000. *Inside City Parks*. Urban Land Institute, Washington D.C.

Harvey, D. 1989. From managerialism to entrepreneurialism: the transformation of urban governance in late capitalism. *Geografiska Annaler* 71:3–17.

Harvey, D. 1996. *Justice, Nature, and the Geography of Difference*. Blackwell, Oxford.

Hay, C. 1995. Re-stating the problem of regulation and re-regulating the local state. *Economy and Society* 24:387–407.

Heynen, N., Perkins, H.A., and Roy, P. 2006a. The political ecology of uneven urban green space: the impact of political economy on race and ethnicity in producing environmental inequality in Milwaukee. *Urban Affairs Review* 42:3–25.

Heynen, N., Kaika, M., and Swyngedouw, E. (eds) 2006b. *In the Nature of Cities*. Routledge, New York.

Hobson, K. 2006. Enacting environmental justice in Singapore: performative justice and the Green Volunteer Network. *Geoforum* 37:671–681.

Holston, J. 1999. Spaces of insurgent citizenship. In: J. Holston (ed.) *Cities and Citizenship*. Duke University Press, Durham.

Houston, D. and Pulido, L. 2002. The work of performativity: staging social justice at the University of Southern California. *Environment and Planning D* 20:401–424.

Jessop, B. 1995. The regulation approach, governance and post-Fordism: alternative perspectives on economic and political change? *Economy and Society* 24:307–333.

Jones, P.S. 2003. Urban regeneration's poisoned chalice: is there an impasse in (community) participation-based policy? *Urban Studies* 40:581–601.

Jones, R.E. and Carter, L.F. 1994. Concern for the environment among black Americans: an assessment of common assumptions. *Social Science Quarterly* 75:560–579.

Jordan, W.R. 1994. "Sunflower forest": ecological restoration as the basis for a new environmental paradigm. In: A.D. Baldwin, J. de Luce, and C. Pletsch (eds) *Beyond Preservation: Restoring and Inventing Landscapes*, pp. 17–34. University of Minnesota Press, St. Paul.

Kearns, A.J. 1992. Active citizenship and urban governance. *Transactions of the Institute of British Geographers* 17:20–34.

Keil, R. and Graham, J. 1998. Reasserting nature: constructing urban environments after Fordism. In: B. Braun and N. Castree (eds) *Remaking Reality*. Routledge, New York.

Koehler, D. and Wrightson, M.T. 1987. Inequality in the delivery of urban services: a reconsideration of the Chicago parks. *The Journal of Politics* 49:80–99.

Light, A. 2003. Urban ecological citizenship. *Journal of Social Philosophy* 34: 44–63.

Lind, A. 1997. Gender, development and urban social change: women's community action in global cities. *World Development* 25:1205–1223.

Mayer, M. 1995. Urban governance in the post-fordist city. In: P. Healy, S. Cameron, S. Davoudi, S. Graham, and A. Madanipour (eds) *Managing Cities: the New Urban Text*, pp. 231–249. Wiley, New York.

McInroy, N. 2000. Urban regeneration and public space: the story of an urban park. *Space and Polity* 4:23–40.

Miles, I., Sullivan, W.C., and Kuo, F. 1998. Ecological restoration volunteers: the benefits of participation. *Urban Ecosystems* 2:27–41.

Mitchell, D. 2003. *The Right to the City: Social Justice and the Fight for Public Space*. Guilford, New York.

Mohai, P. and Bryant, B. 1998. Is there a "race" effect on concern for environmental quality? *Public Opinion Quarterly* 62:475–505.

Pain, R. 1997. Social geographies of women's fear of crime. *Transactions of the Institute of British Geographers*. 22:231–244.

Pain, R. 2001. Gender, race, age and fear in the city. *Urban Studies* 38:899–913.

Pincetl, S. and Gearin, E. 2005. The reinvention of public green space. *Urban Geography* 26:365–384.

Rabrenovic, G. 1995. Women and collective action in urban neighborhoods. In: J.A. Garber and R.S. Turner (eds) *Gender in Urban Research*, pp. 77–96. Sage, Thousand Oaks.

Raco, M. and Imrie, R. 2000. Governmentality and rights and responsibilities in urban policy. *Environment and Planning A* 32:2187–2204.

Robbins, P. 2004. *Political Ecology: a Critical Introduction*. Blackwell, Oxford.

Rocheleau, D., Thomas-Slayter, B., and Wangari, E. (eds) 1996. *Feminist Political Ecology*. Routledge, New York.

Rose, N. 1999. *Powers of Freedom: Reframing Political Thought*. Cambridge University Press, Cambridge.

Ruddick, S. 1996. Constructing difference in public space: race, class, gender as interlocking systems. *Urban Geography* 17:132–151.

Schroeder, H.W. 2000. The restoration experience: volunteers' motives, values, and concepts of nature. In: P. Gobster and B. Hull (eds) *Restoring Nature: Perspectives from the Social Sciences and Humanities*. Island Press, Washington D.C.

Schroeder, R.A. 1999. *Shady Practices: Agroforestry and Gender Politics in the Gambia*. University of California Press, Berkeley.

Scott, J.C. 1990. *Domination and the Arts of Resistance*. Yale University Press, New Haven.

Swyngedouw, E.A. and Heynen, N.C. 2003. Urban political ecology, justice and the politics of scale. *Antipode* 35:898–918.

Taylor, M. 2007. Community participation in the real world: opportunities and pitfalls in new governance spaces. *Urban Studies* 44:297–317.

Thompson, C.W. 2002. Urban open space in the 21st century. *Landscape and Urban Planning* 60:59–72.

Tickell, A. and Peck, J.A. 1992. Accumulation, regulation and the missing geographies of post-fordism: missing links in regulationist research. *Progress in Human Geography* 16:190–218.

Tickell, A. and Peck, J.A. 1995. Social regulation after fordism: regulation theory, neo-liberalism and the global-local nexus. *Economy and Society* 24:357–386.

Valentine, G. 1989. The geography of women's fear. *Area* 21:385–390.

Wekerle, G.R. 1999. Gender planning as insurgent citizenship: stories from Toronto. *Plurimondi* 1:105–126.

Wekerle, G.R. 2005. Domesticating the neoliberal city: invisible genders and the politics of place. In: W. Harcourt and A. Escobar (eds) *Women and the Politics of Place*, pp. 86–99. Kumarian Press, Bloomfield.

Whitehead, M. 2003. (re)Analysing the sustainable city: nature, urbanization and the regulation of socio-environmental relations in the UK. *Urban Studies* 40:1183–1206.

Wolch, J.R. 1990. *The Shadow State: Government and Voluntary Sector in Transition*. The Foundation Center, New York.

Wolch, J., Wilson, J.P., and Fehrenbach, J. 2005. Parks and park funding in Los Angeles: an equity-mapping analysis. *Urban Geography* 26:4–35.

Young, I.M. 1990. *Justice, Difference, and the Politics of Identity*. Princeton University Press, Princeton.

Chapter 9

Rust-to-resilience: Local Responses to Urban Vulnerabilities in Utica, New York

Jessica K. Graybill[1]

Introduction: Placing Vulnerability and Resilience Studies in the City

Scholarship on vulnerability – and resilience in multiple disciplines provides poses different models and frameworks for these concepts situated in different socio-natural communities. Vulnerability and resilience studies are open to multiple interpretations, and may even represent a "conceptual cluster" of frameworks, ideas, and applications advocating different conceptualizations of what they are. Natural scientists and engineers might apply vulnerability to describe or analyze vulnerable or resilient circumstances or phenomena. Social scientists mighty utilize inductive or deductive approaches to vulnerability to explain or describe circumstances or phenomena. Policy and decision makers might attempt to define, theorize and apply vulnerability understandings to help prioritize policy options for specific circumstances. Cutter (2003) calls for vulnerability science today to "require trans disciplinary linkages, methodological pluralism, place-based knowledge, and a continued practical focus on policy relevancy" (2003, p. 8).

Resilience is also variably conceptualized across and within disciplines. For example, within ecology, two definitions prevail, both related to understanding ecosystem changes but with different understandings of the scales of change (Holling et al., 1995). First, resilience can measure how ecosystems maintain stability in "known" steady states and how quickly they return to their equilibrium points after disturbance. This traditional conceptualization of resilience – engineering resilience – assumes that knowledge of the steady states, magnitudes of disturbances, and causes and effects of disturbances are knowable. Applied to urban ecosystems, this understanding may lead to the assumption that cities are manageable and futures are predictable by studying and managing (e.g., planning for or excluding) perturbations. Second, resilience can mean how disturbances can create conditions for changing equilibrium states altogether, suggesting that there

---

1 I gratefully acknowledge the Upstate Institute and Geography Department at Colgate University for their long-term support of my research in Utica. I deeply thank my students, who have committed themselves to assisting me with this work with their boundless enthusiasm and time: Andrew Colbert, Katrina Engelsted, Sarah Hesler, Tara James, Christine Kana, Swetha Peteru, and David Pokorny.

is no return to known steady states. In this version, resilient systems are complex, non-linear, and self-organizing and the existence of steady states is inherently questioned (Berkes and Folke, 2000; Gunderson and Holling, 2002).

Applied to urban settings, the second conceptualization suggests that uncertainty and multi-equilibria exist across time and space and urban ecosystems may exhibit robustness under changing socio-natural conditions. Developing greater understanding of urban socio-natural systems, then, requires historical and spatial knowledge of cities, often best gained locally. Berkes and Folke's (2000) ask whether "resource management be improved by supplementing scientific data with local and traditional knowledge?" (2000, p.13). They see traditional resource management as a local endeavor and local knowledge of how socio-natural systems form, function, and reform after a disturbance as an untapped source of knowledge and experience. Indeed, they "assume that every society has its own means and adaptations to deal with its natural environment, its own *cultural capital*" (emphasis in original; 2000, p.13). Cities are thus underexplored local ecosystems with local knowledge and management systems that are traditionally overlooked when addressing socio-natural systems and management regimes as they change over time and space.

Nature–society interactions (variably called social–ecological systems, socio-nature, or coupled human and natural systems – CHANs – depending on disciplinary context) are created and managed differently across the spectrum of human-nature coupling (e.g., urban to rural) with different temporal and spatial scales of urgency. Nothing emphasizes this point more than urgent scholarly and management trajectories related to urbanization and climate change. For example, a megacity located in a tropical coastal region of a lesser developed country will likely experience the socio-natural hazards of rapid urbanization and sea level rise very differently than a rural village in the inland temperate north of a more developed country. As a scholar, I hesitate to place hierarchical importance on "natural" or "social" issues that are associated with vulnerability or resilience in any setting, but a hierarchy of importance exists for *people* living in *places* that emphasizes the vulnerability of social or natural systems at different temporal or spatial scales. For example, the coastal megacity in a developing country may find it timelier to manage sea-level rise than to alleviate crushing poverty in slums, both of which have implications for the socio-natural setting of the city.

Creating relevant management scenarios for urban ecosystems demands local knowledge, which provides invaluable understandings of urban forms and functions. Berkes and Folke (2000) advocate incorporating Traditional Ecological Knowledge (TEK) in analyses of local ecosystems. Here, I adapt the term TEK, reconceptualizing it as Urban and Local Ecological Knowledge (ULEK), for application to urban settings to analyze how change, vulnerabilities, and resistance are tackled locally in cities. Tapping into ULEK provides new understandings of socio-natural vulnerability and resilience in one understudied ecosystem, the city.

Application of ecosystem concepts to urban areas is not new, and scholarship linking ecology and the social sciences concerned with urban regions began the early 20th century (e.g., the Chicago School; see Park and Burgess, 1925) and is international in scope (e.g., the Berlin, Australia and Seattle schools; see Sukopp et al., 1995; McDonnell et al., 2009 and Head and Muir, 2007; and Marzluff et al., 2008 and Alberti, 2009, respectively). Today's urban ecology is marked by the re-emergence of nature and society as indivisible counterparts. Humans and cities are part of nature and must be considered as such if we are to understand the ecological challenges confronting urban(izing) and global(izing) societies. Challenges for the urban ecologist in the 21st century are both "natural" and "human," meaning that they address the world outside human bodies and societies (e.g., natural landscape on which cities are imprinted and in which humans settle) and the world attributable to humans (e.g., culture, society, environmental concern).

In this chapter, I conceptualize vulnerability as a condition and state of being actively produced by social and ecological factors that may:

1. deepen over time for individuals or communities;
2. be leveraged for multiple reasons when mitigation or adaptation is pursued at multiple scales; and
3. potentially diminish through social and policy decisions made at local to global scales.

Local knowledge about cities, and the knowledge local residents have about their city, can identify, describe, and anticipate or mitigate who or what becomes or stays vulnerable or resilient. Local actions and knowledge, then, contribute to defining the socio-natural system, creating the relationships within the system, and providing understandings of past – and potentially future – change. I apply Berkes and Folke's (2000) understanding of the importance of cultural capital in addressing the management and development of people and places in an urban ecosystem by examining how local forces manage and maintain social and natural resources. I use the term resilience to describe individual or societal resistances to vulnerability in and of the city,[2] addressing both the causes and outcomes of vulnerability. For social and ecological systems, resilience is a measure of: (1) the magnitude of stress absorbed while maintaining functions, and (2) their ability to self-organize and learn (Folke et al., 2001; Turner et al., 2003; Fraser et al., 2003). As suggested by Folke et al., (2002), resilient socio-natural systems are diverse and flexible, absorbing shocks without change in fundamental system parameters. To find ways to build and enhance resilience and hence guide socio-nature along more sustainable trajectories is a central objective for sustainability science (Folke et al., 2002; Kates et al., 2001). Resilience can thus be understood as adaptation to

---

2   Vulnerability *in* the city means the fragile conditions or exposure (e.g., to hazards or change) that individuals or nature might experience within part of the urban setting due to anthropogenic or natural factors. Vulnerability *of* the city means the fragile conditions or exposure of the city itself to anthropogenic or natural factors. Vulnerability in vs. of marks a scalar difference in conceptualizing vulnerabilities within an urban ecosystem.

social or natural changes, but not all resilience is necessarily adaptation (a lack of adaptation may prove to be resilience, too).

Later, I explore vulnerability and resilience in and of the city through the case study of Utica, New York. Utica is a declining city located in North America's northeastern "rust belt" region. From a long period of industrialization and growth, Utica has well-developed infrastructure and governance, but ~40 years of sustained out-migration in the post-industrial era has left the city with a polluted environmental legacy, a declining tax base, and decaying built environment. Stemming further decline is a main concern for the city, which views its future as providing settlement for refugees. Using Utica as my case, I describe and discuss the meanings and implications of how vulnerability – and resilience – are simultaneously managed, maintained, and created.

My framework for analyzing Utica's socio-ecological system is developed after Berkes and Folke (2000), but adapted for urban settings. Figure 9.1 conceptualizes components of Utica's socio-natural urban system as they relate to the concepts of vulnerability and resilience. The conceptual framework provides a visualization of the origins and drivers of current-day change in Utica, identifying physical and social patterns and processes and, potentially, outcomes associated with change. Elements in this diagram are ones I have identified as key to understanding Utica's socio-ecological system – there are certainly others to be considered. The crux of this diagram, however, is not the individual elements, but the varied interactions

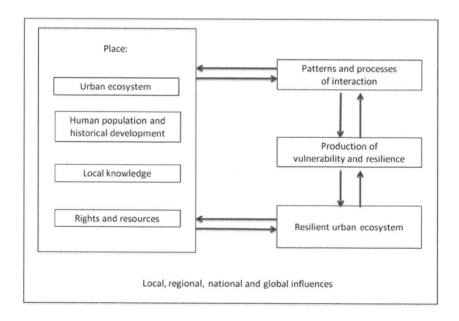

Figure 9.1   A conceptual diagram of Utica's urban ecosystem as it relates
            to the concepts of vulnerability and resilience

among them in space and time that operate to create, maintain, or address vulnerabilities and resilience in and of the city.

Four elements describe the elements and connections of Utica's socio-ecological system. First, the urban ecosystem has biological and physical components that rationalize Utica's existence historically. Recognition of the ecosystem upon which the city was founded explains the development of Utica's complex and diverse socio-ecological system. Second, human population and industrial development is an important analytical element at the community level that describes changing characteristics of the urban population changed over time and space. By understanding how people use resources and space over time as industrial development has changed, the impact of human settlement patterns and economic production on the outwardly expanding city becomes better known. Third, emphasis on local knowledge underscores the importance of humans as agents of influence in this landscape and the socio-natural knowledge they acquire in order to address threats and create opportunities for the future. This is crucial for understanding urban ecosystem resilience: how do local people organize and use knowledge of the city to thrive? What kinds of local environmental knowledge do people acquire and use? How might the application of local environmental knowledge in making decisions impact the city's biophysical functioning? Finally, rights and resources refer to how the city's institutions manage and maintain this urban ecosystem. For example, how do local institutions use natural or social resources in times of scarcity or abundance in the urban ecosystem? Finally, it is crucial to ask how do all of these elements interact through time to influence the capacity of the city to respond to threats and to capitalize on strengths for managing future urban socio-ecological relationships. I aim to show – and further question – how local knowledge and management practices in Utica have built on ULEK to cope with dynamic socio-natural change. Understanding the vulnerabilities and resilience created by urban change is directly related to understanding the patterns and processes of interactions, and no discussion of urban ecosystems is complete without considering the actions and changing phenomena that produce outcomes. In Figure 9.1, I recognize that vulnerabilities and resilience may be starting points for new or altered urban patterns and processes and thus do not link outcomes directly to resilient urban ecosystems (although the connection is suggested).

These conceptual frameworks are neither theory nor models. Theories seek to provide explanation of phenomena based on observation, experiments and reasoning, often aimed at prediction. Models represent systems, allowing for investigation of individual aspects and sometimes for predicting future outcomes. Alternatively, conceptual frameworks provide ways of thinking about and making sense of urban ecosystem phenomena that may reveal patterns of activity and interactions through understandings of processes. This is key in developing an understanding of the complexity of urban ecosystems and then applying the concepts of vulnerability and resilience to them.

Utica, NY: A Brief History

Conceptualizing Utica's resilience in the 21st century requires understanding the city's history. I review the city's history as related to local environment and historic urban development, identifying periods of growth and decline that affect the city's socio-ecological composition.[3] Figure 9.2 provides an overview of population change in Utica since 1820, valuable for understanding Utica's historical growth and decline.

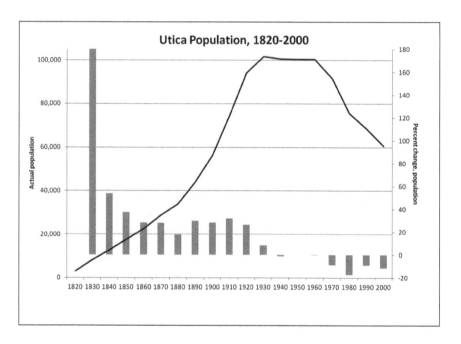

Figure 9.2   Population change in Utica, NY (data source: US Bureau of the Census, 2000)

Crossroads Era (until 1817)

Utica lies along a shallow and slow part of the Mohawk River that made for a protected and easily accessible meeting place for regional Iroquois groups in upstate New York. The first European explorers knew it as a thriving Native American trading site and early European mercantilism also developed here near

---

3   Excellent resources about Utica's historical development include Noble (1999) and Cardarelli (2009). Explorations of the changing ethnic composition include Kraly and Van Valkenburg (2002), Pula (2002), and Owens-Manley and Coughlan (2005).

the site of Fort Schuyler (c. 1773; Cookinham, 1912). European settlement here and in other small settlements throughout the region initiated further growth and eventual development of a transportation corridor – what was to become State or Genesee Street – through the region. Additional roads (e.g., the Seneca Turnpike), became important for continued urban development and Utica became a transportation hub for consumer goods moving from western New York to the Great Lakes region (Oneida County Historical Society).

Spatially, this era marked changing types of crossroads for travelers and traders. Regionally, populations changed from dominantly Native American to dominantly European, marking the rise of white settlements across this landscape. Regional growth contributed to the establishment and expansion of the early settlement into a village.

Canal and Manufacturing Era (1817–1970)

Regional canal development linked people and goods in new ways, and cities like Utica became crucial for transportation and infrastructure development. A main driver of Utica's growth, the Erie Canal (Bernstein, 2005), boosted population first in response to needing engineers and skilled workers and later as a resting point: Utica was regarded as halfway between Albany and Syracuse. As the city grew, so did spaces designated for urban nature (i.e., parks), donated to the city by local industrialists for aesthetic and leisure reasons (ongoing unpublished research by the author). By the late 19th century, established greenspaces dotted the city, most notably in elite south eastern Utica. Olmstedian in design, the parks from this era are reminiscent of the Emerald Necklace in Boston, providing a chain of greenspaces linked by corridors accessible to all (see Figure 9.3, Eras 1 and 2 parks). Roscoe Conklin and Proctor parks remain gems in Utica's greenspace network today, which extends citywide as leisure and recreational parks, medians, and grassy areas near major roadways.

Light industry (textiles, tool and die) thrived in Utica because of the physical geography along the Mohawk River. Shallow waters incapable of deep-water energy creation eliminated the possibility of heavy industry development with technologies of the era. The city instead developed textile mills and by the 1840s, steam-run mills dominated Utica's manufacturing. This time period, peaking in 1918, is referred to as the "Textile Era," and thousands were employed, making Utica the "knit goods capital of the world" (Greene, 1925)[4] until the 1950s. The tool and die industry remained into the 1970s, but eventual industrial migration to the American South and abroad marked the beginning of Utica's declining, rustbelt character century (Oneida County Historical Society).

---

4   According to Greene 1925, 1920s Utica maintained 370 factories employing 18,564 people valued annually at $78,000,000. As the US' main textile center, Utica produced many items associated with the garment industry. In light industry, Utica produced multiple metal-based commercial and retail items.

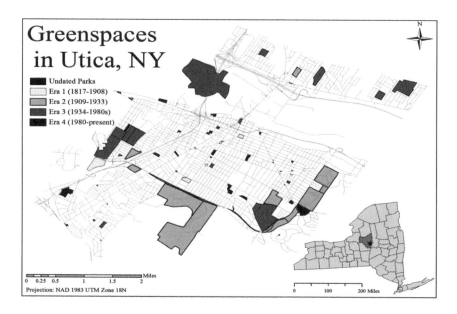

**Figure 9.3   Utica's public greenspaces as of 2010. Here they are categorized
by eras of development**

Spatially and demographically, Utica grew throughout the canal and
manufacturing era until 1940, when minor decline began (~1 percent). Initially
marked by increasing urban population density, this growth period concluded in
suburban and regional sprawl. Suburban and regional sprawl are distinguished
by proximity to or distance from the city, respectively. Because textile mills were
located in Utica and in other towns along the Erie and Chenango canals, regional
identity and competition among mills and towns grew. Thus, people who might
have worked in Utica's mills were spread out, but thought of Utica as a regional
labor and skill hub. In the mid-20th century, suburban sprawl to adjacent villages-
turned-cities (e.g., New York Mills, New Hartford) signified urban economic
growth, but also portended urban decline.

Rustbelt Era (1970–present)

Flight of most manufacturers to other regions occurred by the 1970s, but Utica's
manufacturing era definitively ended when The Chicago Pneumatic Company closed
in 1997 <www.oneidacountyhistory.org>. Utica's rust belt era is marked, as in other
northeastern US rust belt areas as decline of the (a) importance of waterways for
manufacturing and (b) canals and railways as transportation corridors (see Cooke,
1995). By the late 20th century, emphasis on local–regional industrial corridors was
outdated, as globally outsourced industry and new modes of transportation linked

people with more desirable places for living and working (e.g., suburbia, exurban farms, growing cities). Abandoned industrial sites, many classified as brownfields and/or superfund sites by the Environmental Protection Agency, dot the city and are in various states of remediation[5] (Environment DEC, 2008).

Linked to manufacturing decline, population decline began in 1940 (1 percent), became precipitous by 1970 (9 percent), roughly doubled by 1980 (17.4 percent), decreased slightly by 1990 (9.2 percent) and increased again by 2000 (11.6 percent). The 2000 census reported 60,651 residents, a 39.6 percent decrease from 1960. In short, 40 years of population decline is poignant when placed in regional context: seven of the 65 lowest performing US cities are in New York State, and most of these are along the historic canal and manufacturing corridors (Vey, 2007).

Spatially, sprawl into surrounding villages and rural areas in search of better socio-natural environments for families and investment caused urban decay in Utica, as abandoned houses decayed in or were removed from neighborhoods. As in other rustbelt cities and their suburbs, the inner city declined while suburbia prospered. Maintained with ever-diminishing fiscal resources, the urban greenspaces network and park infrastructure (e.g., restrooms, community centers) remains outdated and only minimally maintained (i.e., mowing) in high use and high visibility areas for lack of funding. In 2007, Utica's Parks Commissioner remarked that a master plan to re-conceptualize the parks system to meet the needs and desires of today's citizens is "not in the budget" (D. Short, personal communication).

"Second Chance City" Era (1980–present)

In 1978, one family brought the first Vietnamese refugees to Utica (McGill, 2007). With aid from a Catholic charity and partnership with a resettlement agency, the nonprofit organization Mohawk Valley Resource Center for Refugees (MVRCR; <www.mvrcr.org>) was founded. Since inception, the MVRCR has settled over 13,500 refugees from at least 34 different countries in Utica,[6] giving rise to Utica's nickname, "Second Chance City." Figure 9.4 shows the top four countries of origin of former refugees resettled to Utica. It also indicates decreasing resettlement over time, attributable to national policy changes regarding incoming refugees after 9/11 (E. Kraly, personal communication). Additionally, resettlement from different countries of origin has changed over time, shifting from dominantly Bosnian

---

5   According to the US Environmental Protection Agency, brownfields are "real property, the expansion, redevelopment, or reuse of which may be complicated by the presence or potential presence of a hazardous substance, pollutant, or contaminant" <http://epa.gov/brownfields/overview/glossary.htm>.

6   By late 2010, so many former refugees have resettled, renovated existing housing stock, and stayed in Utica that the MVRCR seeks resettlement options in Rome, New York (20 miles away) due to the lack of affordable housing available for incoming refugees (E. Kraly, personal communication).

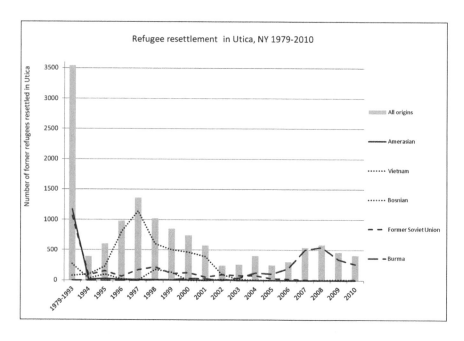

Figure 9.4   Chart showing the top four resettled refugee groups by number resettled against a background of all resettled people. Data source: Mohawk Valley Refugee Resource Center, 2010

resettlement to dominantly Burmese resettlement recently. These changes reflect shifting (a) conditions in countries of origin and (b) decision-making processes by the State Department about refugees resettlement (see Singer and Wilson, 2006). Since 1978, the number of people resettled has changed eased over time (mainly in response to non-local forces) but by the 1980s, resettlement acts as a bandage stemming the population outflow from Utica. Also important to recognize is that refugees in the US are tied to their resettlement location for at least five years. This impacts the city, providing Utica with a relatively stable population increase in the short term and, perhaps, in the long term.

Spatially, influx of refugees means inner city revitalization. Residing in Utica for five years and, in many cases, beyond, former refugees tend to settle in the city, buying and renovating rundown, affordable properties (Fulton, 2005; Chanatry, 2006). Housing values have increased since refugees have reclaimed the city and as residential neighborhoods revitalize, so do family-owned businesses. Bleeker Street, once a thriving Italian–American neighborhood shopping area, declined in the late 20th century but is revived today with the infusion of ethnic-oriented businesses. In one city block, Bleeker Street's groceries, restaurants, and beauty salons target a diverse population. At least nine different ethnic groups are represented in the ownership, products or languages spoken or advertised there today, reflecting the diversity of former refugees in the city. The trend of decline in Utica has only begun to turn around in the 21st century with such urban reinvestment and other initiatives, such as restoration of iconic buildings like The Stanley Theater and Hotel Utica.

Beginning around 2002, interest in creating community greenspaces has led to community garden initiatives, intent upon bringing together urban residents around nature in the city (e.g., see <www.forthegoodinc.org>). Non-profit organizations see community gardens as potentially sites for bringing citizens together to enjoy nature in the city and to develop common interests from diverse cultural backgrounds. A survey conducted in summer 2009 by students at Colgate University indicates mixed results from community garden development (see <https://sites.google.com:443/a/colgate.edu/jkgraybill/>). Residents near community gardens were (a) unaware of their presence or purpose or (b) not interested in community gardening. The survey also revealed that information about the gardens was not provided in their native languages for immigrant populations, so access to knowledge about them was limited. A tour of community gardens in September 2010 showed that while many had been planted, few people visited them and harvests went unreaped, questioning whether garden greenspace is important for Utica.

Exploring Urban Vulnerability and Resilience in Utica

Historical analysis of Utica's growth and decline, noting changing social and natural components, provides perspective for evaluating the vulnerability(ies) and resilience(s) in and of the city today. Below, I address two main points addressing Utica's vulnerability and resilience. First, I find three major factors make Utica vulnerable and resilient simultaneously. Second, I argue that these factors cannot be understood independently and must be examined holistically. The three main factors creating Utica's simultaneous vulnerability and resilience are: (1) its post-industrial status, (2) ongoing yet changing influx of former refugees, and (3) the presence of a well-developed suburban and exurban landscape. I discuss Utica's vulnerability and resilience separately, as this is helpful in an initial discussion of them. However, I also recognize that some vulnerabilities may create resilience, and vice versa.

Vulnerable Utica

From the socio-ecological history of Utica and the understanding of vulnerability developed in this chapter, I identify and briefly describe below six major factors that indicate vulnerability in and of Utica.

Post-industrial built environment Empty or partially repurposed industrial buildings dot Utica's landscape, as do brownfields and vacant lots covered in asphalt or urban prairie.[7] Most industry in Utica remains located in or along certain sections and corridors of the city, meaning that brownfields and vacant industrial sites are often spatially connected, uninterrupted by residential space. However, residential neighborhoods are close (i.e., less than one city block away) to many industrial sites and corridors. The Bossert Manufacturing Site, e.g., is a reclaimed brownfield resulting from a former sheet metal industrial complex and is located on a major arterial route and backs up against a residential neighborhood. The City of Utica owns the site but is unable able to sell it. Residential property values are lower here than in other similar neighborhoods and vacancies turn into blighted, boarded up homes or urban prairie.

Declining population Utica's 40 percent population decline is situated in overall decline of many northeastern and mid western US cities (Hobbs and Stoop, 2002), but sustained and ongoing decline in Utica deserves further spatial explanation. First, Utica's loss of manufacturing jobs in the mid-late 20th century led to out-migration of skilled workers. Second, as in other cities, suburban growth around Utica also caused urban population decline (Jackson, 1985). Suburban growth does not necessarily remove people from working in the city, but it does remove them from the necessity of living in the city. Indeed, Utica's suburban population is healthy, and review of Census 2000 data indicates that suburban population decline has not been as precipitous as in the city.

Out-migration of the creative class The creative class – an emergent class composed of knowledge workers, intellectuals, and artists (Florida, 2002) – is an important economic driver in the post-industrial era. The creative class does not pursue traditional labor (e.g., industry, agriculture) and instead includes "people in design, education, arts, music and entertainment, whose economic function is to create new ideas, new technology and/or creative content" (8). The Public Policy Institute (2004) notes migration of highly educated and creative people away from Utica today, and write that over time and space, out-migration of the creative class leads to fewer possibilities for upward mobility and interesting, non-service sector employment in post-industrial places. The creative class thrives in growing cities, suburbs and rural fringes. Because suburbs provide services and lifestyles desired by much of the American public, it is in the suburbs where the creative class largely

---

7   While usage varies, urban prairie describes vacant urban lots reverted to green spaces. Resulting from demolition of environmentally or socially harmful spaces (e.g., defunct manufacturing buildings, blighted residential structures), urban prairies are sometimes planned by local governments to facilitate community green spaces (such as community gardens, wildlife habitats), but more often result from a lack of coordinated redevelopment planning, resulting in patchworks of unmanaged and unintended urban green space (Whirter, 2001).

enriches residential neighborhoods, greenspaces, and educational opportunities. Throughout the 20th century, historically urban businesses relocated to suburban and rural regions to capitalize on lower rents and new markets (Jackson, 1985). In Utica, residential and commercial suburbia sprawls into rural farmland, providing affordable havens for those who need proximity to the city but desire locations with more land and different possibilities. In the 21st century, well established suburban and exurban communities near Utica draw and keep creativity out of the city, at home and at work.

Taxes Manufacturing collapse and large out-migration has left Utica with a declining tax base since the mid-20th century. Coupled with overall high taxes in New York State (among the highest in the nation; see Public Policy Institute, 2004), individuals and companies are loath to moving into places like Utica, causing further stagnation and decline. For urban nature, it has meant that "recreational" nature (e.g., parks) is maintained only at a minimum level and "ecological" nature (e.g., urban prairie) is not managed, maintained or monitored.

Former refugee resettlement trends Twenty years of former refugee resettlement has stemmed some of Utica's population decline, but it is unknown if population stabilization through refugee resettlement will continue because of changes in local and federal programs. For example, review of refugee resettlement after 9/11 caused much questioning of the US' role in resettlement processes. It is unknown and understudied whether refugees remain in initial resettlement locations after five years or if they leave for other communities near Utica (suburban or rural) or in other US locations (E. Kraly, personal communication).

Culture clash Historically a city of immigrants, with notable Italian, Polish, Welsh, and African American communities, the arrival of former refugees from other locations (see Figure 9.4) challenges Utica to promote tolerance and education about global cultures and languages. Transition and re-imagining Utica's cultural identity is not always smooth, especially in ethnic neighborhoods where ideas about "turf" and cultural identities are long established. For example, the historically African American Corn Hill neighborhood is becoming a multi-cultural neighborhood with increasing ownership and rent by former refugees (most notably by Bosnians), leading to feelings of displacement among older communities and ethnically based rivalry among younger generations.

In Utica, vulnerabilities are social and ecological and result from both existing conditions and are created by processes of change. Human vulnerabilities due to post-industrial decay of the older built environment include declining neighborhood attractiveness and property values (Chanatry, 2006), soil and water contamination from former industrial sites (Ghosh et al., 2003), and lingering lead paint and asbestos in older houses (Korfmacher and Koholski, 2008). Long-term residents may not care about appearances or heed warnings about harmful substances. Newer residents may not have relocation choices or aren't aware of harmful substances in the built environment. Additionally, continued out-migration reduces the city's tax base, depleting the city's ability to address environmental pollution or it's environmentally compromised and blighted image. The only migration into the city by former refugees is not by choice, but by coordinated city and federal government policy and efforts. Vulnerabilities from this type of in-migration include an increased initial tax burden to the city from new low-income residents (Fulton, 2005) and the potential for cultural or social misunderstanding among urban residents. Additional vulnerabilities in and of the city are due to uncertainty whether (1) former refugees will remain in Utica, creating a return on the city's investment in refugees, and (2) federal support for resettlement programs will continue.

Resilient Utica

Despite the vulnerabilities discussed above, Utica maintains revitalization potential because of social resistance to decline. Indeed, recent renovation downtown indicates desire and ability to create a strengthened urban core. Additionally, several factors noted to make Utica vulnerable could also be understood as producing resilience. Below I describe three factors creating resilience in Utica's socio-ecological system.

In-migration Influx of multiple ethnicities brings new ideas, skills, and cultures into the city. While the lack of knowledge of English by incoming former refugees could be understood as a potential vulnerable condition for individuals, the MVRCR and other local organizations provide English as a second language training and job placement for former refugees, thus alleviating the state of being vulnerable. This indicates that communities in Utica exhibit social resilience by adapting to changing local conditions by creating better transitions for all local citizens as the city gains a global identity.

Utica's labor force is renewed as former refugees have strong personal incentives to acquire jobs quickly after gaining rights to work in the US. In addition to providing higher skilled labor, they occupy low-paying, low-skilled jobs that entrenched Americans will not take (Coughlan and Owens-Manley, 2006). Providing an economic boost to the city, the workforce is re-energized, creating resilience for individuals and communities and overall resistance of the city to further economic decline.

In-migration revitalizes older urban neighborhoods as surplus housing is renovated, especially in lower income older neighborhoods where larger homes from the turn of the last century languish in varying states of decay. Former refugees – especially the established Bosnian and Russian communities – revitalize individual houses by refurbishing single- and multi-family homes. Fulton (2005) notes that former Bosnian refugees bought many houses for only tens of thousands of dollars, renovating them and changing the ethnic composition and derelict nature of entire neighborhoods. In this manner, the urban environment is changed, the city tax base increases and property values rise, all indicating resilience in and of the city of Utica. Here, resilience is related to enhancing capacity, representing engineering resilience, because the social system is returning to previous levels of functionality through the specific processes of in-migration.

Due to the legal status of former refugees, many local resources are immobile in Utica, at least for the five years that former refugees are required to stay in place. Local urban resources (e.g., labor, income, taxes) remain in Utica for at least that time period and potentially longer, as former refugees build new futures. Ethnic diversity is attractive to many former refugees, and creates space for recovering from shared refugee experiences and for mutual understanding of successfully moving forward in a new urban space. Quite possibly, strong ethnic communities built around former refugee communities, such as the Utica Bosnians, may also attract other people of the same ethnic group from other US resettlement sites. An understudied phenomenon in Utica, there is some anecdotal and statistical evidence that this is occurring. Internal estimates at the MVRCR identify a significant secondary settlement rate, up to several hundred people per year (E. Kraly, personal communication). Additionally, some former refugees become local humanitarian resources for the city or for other incoming refugees, aiding with the resettlement process. This is a way of giving back to that builds community resilience.

Availability of urban space   Availability of multiple kinds of spaces in Utica contribute to its resilience. Plentiful unused space, such as old industrial and manufacturing buildings, vacant lots, and vacant housing, provides potential for urban infill. Repurposing of these spaces, in the forms of community gardens, parks, or second generation commercial or residential lives, requires, imagination and labor, and minimal investment.

I include affordability and access as characteristics of availability, as ability to rent or buy property or space in Utica is key for potential users. For example, the Matt Brewing Company (maker of Saranac beverages) is slowly sparking urban renaissance of local nightlife in West Utica through restaurant and live music rejuvenation. Additionally, influx of Bosnians into Corn Hill and other neighborhoods revives the city's cultural capital and economic base. Affordability of Utica's housing stock brings homeownership opportunities to former refugees, rebuilding pride in and among ethnicities seeking new lives and lifestyles in the US. With this perspective, Utica's decline can be viewed as a positive influence on the ability of newcomers to be able to succeed economically and personally in a new urban environment. Resilience for all local residents is created by redevelopment of neighborhoods and increasing tax bases for entities like local schools. Here, resilience is related to creating new capacity in the urban setting after transition, representing complex, non-steady state resilience, because individuals and the social system create and maintain new kinds of functionality through building lives and spaces anew.

Well-developed suburban and rural surroundings   While preference for suburban-like lifestyles leads people out of cities nationwide (Hobbs and Stoops, 2002), some dollars return to the city from the suburbs (Kivell, 1993). In Utica, well-developed suburbs near the city strengthen urban redevelopment. For example, the arts district in Utica provides suburbanites with dining and entertainment on a "night on the town," something relegated to Utica's past only a few years ago. Additionally, it is well documented in urban geographic literature that while suburban and rural fringe growth removes residents from the city, urban cores such as Utica are still necessary for everyday governmental and financial functioning (Knox and McCarthy, 2005; Kaplan et al., 2009).Because there is a suburban and rural fringe to Utica means that there is a population density near the city and without either of these fringes, Utica may not have made it as a city into the 21st century. Interconnectedness of the regional landscape – city, suburbs, and rural fringe – cause Utica to remain resilient even amidst population decline.

Existing Infrastructure for Urban Nature

Utica's extensive formal parks system (Figure 9.3) includes large tracts of land throughout the city and near ecologically important features, such as rivers and marshes, which act as corridors for many non-human species in the city. Although currently parks management is under-funded and minimally maintained, the presence of relatively intact large swaths of land in and around the city is promising for keeping nature in the city. Many cities today are attempting to bring back or revive nature in the city, but the urban ecosystems around the city's edges – and along the nearby Erie Canal – are relatively intact. While ecology of inner Utica, especially near former industrial sites and corridors, creates vulnerabilities for social and natural systems, nature on the urban edges are assets for future recreational or educational sites, and as functioning ecosystems in and near the city. With continued leadership and vision for urban nature in Utica, the "emerald necklace" that surrounds the city will become a future asset, creating a more resilient socio-natural ecosystem for Utica.

Rust-to-resilience: Utica's Urban Ecosystem

Utica is often portrayed as a blighted, decaying post-industrial rust belt city in the northeastern US. Indeed, it is hard to see anything else when newsflashes about the city refer to cleaning up brownfields, urban decay, and fiscal desperation "solved" with federal support that fills the city with non-English speaking refugee communities, and meager attempts to maintain a city parks network with ever-diminishing resources for an ever-shrinking population. This Utica is indeed a vulnerable urban space, made so by continuing technological innovation that left the period of canals and railroads in history, by American consumers' suburban bias and westward and southward migration, by a legacy of polluting industries that failed to clean up land and waters and the high-tax nature of New York state that cripples new business growth or takeover of older abandoned urban spaces, and by the re-filling of the city with new kinds of people with new desires and wants in their adopted city, but with little power or rights (at least for five years) to transform older social networks in the city as they struggle to learn a new language, culture, and city.

Another way to conceptualize Utica is as a shrinking city that has embraced innovative ways of stemming decline by decision-making and action at the local and federal levels. Issues of urban decay, the declining tax base, and out-migration are countered with hopes for new beginnings for people and for urban spaces as former refugees from around the world find safe haven in the city and its largely welcoming secular and religious communities. Long-derelict urban neighborhoods are beginning to be cleaned up, and the threat of further decay may be decreasing. Increasing tax bases provide initiative for the city to address environmentally polluted spaces in new ways, such as by creating community greenspaces. Should newcomers remain, new generations of Uticans with multiple cultural and social backgrounds will change the image of the city, forcing reevaluation of current understandings of immigrant gateways, multicultural communities, and socio-economic processes in the rustbelt region.

This case study highlights how vulnerability is a process that is actively produced and transformed at individual, community, city, and federal levels. It shows that the specific components involved in the (re)production of vulnerability, and the risks to and exposures of harm from varying things that render people and places vulnerable must be conceptualized differently for different places based on local histories, natural environments, and the needs of people and nature in different places. In short, *geography matters* in conceptualizing and applying the concepts of vulnerability and resilience when researching people and places. In all geographies, there is a hierarchy of importance of social or natural vulnerabilities for people that emphasize different concerns at different temporal or spatial scales. I have suggested that nature–society interactions in Utica's socioecological system are experienced and created variably across the spectrum of human-nature coupling with different temporal and spatial scales of urgency. At the city scale, addressing vulnerability in Utica means stemming human population decline and reclaiming urban spaces from conditions of blight and environmental pollution. At the individual scale, addressing vulnerability in Utica means adapting to local cultural and labor environments to claim citizenship and rights to belonging in a changing urban space.

This case study of Utica suggests that it is impossible to conceptualize vulnerability of an urban socioecological system only. Resilience and its co-conspirators – coping capacities, adaptive strategies, resistance to decline, and recovery after natural hazard or social shock – are inherently part the concept of vulnerability. However, the productions of vulnerability and resilience in Utica are tightly linked. For example, the local attempt to mitigate urban vulnerability due to precipitous population decline in the 20th century was urban renewal through refugee resettlement, subsidized with local, and federal support. At the urban scale, the city solved two vulnerabilities (massive population decline and potentially ending long-term tax base decline) by bringing in new communities necessarily tied to place, thus resisting urban decay and increasing resilience at the city scale. Social and environmental change in Utica is occurring due to refugee dollars and initiative, rebuilding the city in new ways. This is resilience because change to the city brought about social actions, to which even more actions and changes are attributed.

The facilitation of resistance to vulnerability through multi-scalar and collaborative work by individual, city and federal entities is a form of resilience that operates to reinforce and mitigate Utica's vulnerability. The same programs that attempt to stem urban vulnerability may create personal or community vulnerability, as people in the city – newcomers and old-timers –adapt to changing socio-cultural conditions in the city. Thus, residual vulnerabilities remain and new vulnerabilities are created that, in this post-industrial setting, are not alleviated by efforts to become resilient. In addition to social and cultural vulnerability discussed above, vulnerability to environmental pollution remains at the individual and city scales.

In viewing the urban system through the lenses of vulnerability and resilience, it becomes apparent that many questions are raised about these perspectives. For example, in what ways does enhancing a city's capacity for adaptation and the mitigation of harmful effects also place different populations at risk? Are unintended vulnerabilities created in the process of creating resilience to known vulnerabilities? Is creating resilience in existing urban systems an answer to addressing vulnerability, or is more vulnerability created at different scales, in new communities (social or ecological) or in different places? Returning to the conceptual diagram (Figure 9.1), we see how elements of Utica's socio-ecological system interact over time and space, allowing for changing emphasis – temporally and spatially – on the variable factors that create the ever-transforming urban environment. What becomes clear is that vulnerability and resilience are ever-changing and multiscalar phenomena and that complex ecosystems, like the urban ecosystem, must be continually addressed and re-addressed by scholars attempting to pinpoint causes of and solutions for vulnerability or reasons for ongoing resilience. The conceptual framework's flexibility allows for investigating the interrelated notions of vulnerability and resilience because it accommodates systemic change over time. What is important to recognize is that to conceptualize any kind of resilience is to acknowledge *change*, which could be cultural, social, economic, or physical in nature. In this sense, change is a necessary precursor to any kind of resilience. Geographies of vulnerability have never been static, so ways of understanding urban resilience cannot be, either. For example, as a rustbelt-to-resilience city, Utica has survived severe human resource decline and shortages with flexibility and creatively, namely by becoming a refugee resettlement community within the US. This has impacted how resources are available to the city and how individual people gain rights to the city in specific ways.

That Utica's transformation from vulnerable to has been driven by diverse actors and through unlikely collaborations is also noteworthy and demands that conceptualizations of how geographies are formed and maintained or altered over time and space be interrogated in new ways. Resilience – of human systems and their coupled natural systems – is due to communities' responses to vulnerability and change in complex and diverse ways depending on their needs, values, cultures, capacities, institutional forms, and environmental features. Urban resilience, then, is a city's ability to withstand and adapt to socioecological changes over time. What is also clear from the case study of Utica is that vulnerabilities and resilience co-occur, and this is one illustration of the complexities of issues that arise in any study of vulnerability and resilience in the urban setting.

References

Alberti, M. (2009). *Advances in Urban Ecology: Integrating Humans and Ecological Processes in Urban Ecosystems*. Seattle: University of Washington Press.

Berkes, F. and Folke, C. (1998). *Linking Social and Ecological Systems: Management Practices and Social Mechanisms for Building Resilience*. Oxford: Cambridge University.

Bernstein, P. (2005). *Wedding of the Waters*. New York: Norton.

Cardarelli, M.J. (2009). *Dawn to Dusk in Utica, New York*. New Hartford: The Author.

Chanatry, D. (2006). "Utica, N.Y., draws immigrant population." *National Public Radio*. Accessed online August 27, 2010: <www.npr.org/templates/story/story.php?storyid=5182157>.

Comfort, L., Wisner, B., Cutter, S., Pulwarty, R., Hewitt, K., Oliver-Smith, A., Wiener, J., Fordham, M., Peacock, W. and Krimgold, F. 1999. "Reframing disaster policy: the global evolution of vulnerable communities." *Environmental Hazards* 1 39–44.

Cooke, P. (1995). *The Rise of the Rustbelt*. New York: St. Martin's Press.

Cookinham, H.J. (1912). *History of Oneida County N.Y.* Chicago: SJ Clarke Publishing Company.

Coughlan, R. and Owens-Manley, J. (2006). *Bosnian Refugees in America: New Communities, New Cultures*. New York: Springer.

Cutter, S.L. (1995). "The forgotten causalities: women, children, and environmental change." *Global Environmental Change* 5, 181–194.

Cutter, S.L. (2003). "The vulnerability of science and the science of vulnerability." *Annals of the Association of American Geographers* 93, 1–12.

Environment DEC (2008). "Two brownfield cleanups and one site redevelopment announced." New York State Department of Environmental Conservation. Accessed online January 26, 2011: <http://www.dec.ny.gov/environmentdec/47160.html?showprintstyles>.

Florida, R. (2002).*The Rise of the Creative Class: How it's Transforming Work, Leisure, Community and Everyday Life.* New York: Perseus.

Folke, C., Carpenter, S., Elmqvist, T., Gunderson, L., Holling, CS., Walker, B., Bengtsson, J., Berkes, F., Colding, J., Danell, K., Falkenmark, M., Gordon, L., Kasperson, R., Kautsky, N., Kinzig, A., Levin, S., Mäler, K-G., Moberg, F., Ohlsson, L., Olsson, P., Ostrom, E., Reid, W., Rockström, J., Savenije, H. and Svedin, U. (2002). *Resilience and Sustainable Development: Building Adaptive Capacity in a World of Transformations.* ICSU Series on Science for Sustainable Development, No. 3.

Fulton, W. (2005). "Refugee renewal." *Governing* May 2005. Acesssed online December 15, 2010: <http://www.governing.com/topics/health-human-services/Refugee-Renewal.html>.

Ghosh, U. Zimmerman, J.R. and Luthy, R.G. (2003). "PCB and PAH speciation among particle types in contaminated harbor sediments and effects on PAH bioavailability." *Environmental Science and Technology* 37(10) 2209–2217.

Greene, N. (ed.) (1925). *History of the Mohawk Valley: Gateway to the West 1614–1925* (pp. 1823–1855). Chicago: The S.J. Clarke Publishing Company.

Head, L. and Muir, P. (2007). *Backyard: Nature and Culture in Suburban Australia.* Wollongong: University of Wollongong Press.

Hobbs, F. and Stoops, N. (2002). "Demographic trends in the 20th century." U.S. Census Bureau, Census 2000 Special Reports, Series CENSR-4, Washington DC: U.S. Government Printing Office.

Jackson, K. (1985). *Crabgrass Frontier: The Suburbanization of the United States.* Oxford: Oxford University Press.

Kaplan, D., Wheeler, J. and Holloway, S. (2009). *Urban Geography.* Hoboken: Wiley.

Kates, R.W., Clark, W.C., Corell, R. Hall, J.M., Jaeger, C.C., Lowe, I., McCarthy, J.J., Schellnhuber, H.J, Bolin, B., Dickson, N.M., Faucheux, S., Gallopin, G.C., Grübler, A., Huntley, B., Jäger, J., Jodha, N.S., Kasperson, R. E., Mabogunje, A., Matson, P., Mooney, H., Moore, III B., O'Riordan, T. and Svedin, U. (2001). "Sustainability science." *Science* 292 641–642.

Kivell, P. (1993). *Land and the City: Patterns and Processes of Urban Change.* London: Routledge.

Knox, P.L. and McCarthy, L. (2005). *Urbanization: An Introduction to Urban Geography* (2nd edition). Toronto: Pearson.

Korfmacher, K.S. and Kuholski, K. (2008). "Childhood lead poisoning in Oneida County: a needs assessment." University of Rochester, Environmental Health Sciences Center. Accessed online January 30, 2011: <www.envmed.rochester.edu>.

Kraly, E. and Van Valkenburg, K. (2003). "Refugee resettlement in Utica, New York: opportunities and issues for community development." In: J. Frazier (ed.) *Multicultural Geographies* (pp. 125–146). Binghamton: Binghamton University Press.

Liverman, D.A. (2001). "Vulnerability to drought and climate change in Mexico." In: J.X. Kasperson and R.E. Kaspersen (eds) *Global Environmental Risk.* Tokyo: United Nations University Press.

Marzluff, J.M., Shulenberger, E., Endlicher, W., Alberti, M., Bradley, M., Ryan, C., Zum Brunnen, C. and Simon, U. (2008). *Urban Ecology: An International Perspective on the Interaction Between Humans and Nature.* New York: Springer.

McDonnell, M., Hahs, A. and Breuste, J.H. (2009). *Ecology of City and Towns: A Comparative Approach.* Cambridge: Cambridge University Press.

McGill, D. (2007). "The town that loves refugees." *Christianity Today*, February 2007.

Noble, A.G. (1999). *An Ethnic Geography of Early Utica, New York: Time, Space, and Community.* Lewiston: E. Mellen Press.

Oneida County Historical Society (1977). "Utica." In: *The History of Oneida County.* Oneida County.

Owens-Manley, J. and Coughlan, R. (2005). "Adaptations of refugees during cross-cultural transitions: Bosnian refugees in Upstate New York." Accessed January 26, 2011: <http://www.hamilton.edu/Levitt/pdfs/owens-manley_refugee.pdf>.

Public Policy Institute (2004). "Could New York Let Upstate Be Upstate?" Albany: Public Policy Institute. Accessed online September 8, 2010: <http://www.ppinys.org/reports/2004/letupstate.pdf>.

Pula, J.S. (ed.) 2002. *Ethnic Utica.* Utica: Ethnic Heritage Studies Center at Utica College.

Singer, A. and Wilson, J.H. (2006). "From 'there' to 'here': refugee resettlement in metropolitan America." The Brookings Institution Living Cities Census Series. Accessed online November 1, 2011: <http://www.brookings.edu/reports/2006/09demographics_singer.aspx>.

Sukopp, H., Numata, M. and Huber, A. (eds) *Urban Ecology as the Basis of Urban Planning.* The Hague: SPB Academic Publishing.

Turner, B.L., Kasperson, R.E., Matson, P.A., McCarthy, J.J., Corell, R.W., Christensen, L., Eckely, N., Kasperson, J.X., Luers, A., Martello, M.L., Polsky, C., Pulsipher, A. and Schiller, A. (2003). "A framework for vulnerability analysis in sustainability." *Proceedings of the National Academcy of Sciences* 100(14), 8074–8079.

Vey, J.S. (2007). *Restoring Prosperity: The State Role in Revitalizing America's Older Industrial Cities.* Brookings Institution Metropolitan Policy Program.

Whirter Cameron M.C. (2001). "Broken Detroit: death of a city block." Detroit News. 5-day series (consecutive), beginning Sunday June 17, 2001.

Chapter 10

The Privilege of Staying Dry: The Impact of Flooding and Racism on the Emergence of the "Mexican" Ghetto in Austin's Low-Eastside, 1880–1935

Eliot M. Tretter and Melissa Adams

From a Great Flood – A "Negro" Spiritual
I heard the people saying up in the tree,
Come somebody and save poor me,
From a great flood.
There was a great flood on the Colorado Stream,
The Cattle was lowing, the children was crying,
Mother was saying my friends are dying,
From a great flood

Franklin, 1935

Environmental racism often refers to ways in which non-white minorities are disproportionately exposed to human–created hazards and has often overlooked how the exposure to natural hazards is also unevenly borne by members of society. Examining a case study of flooding in Austin, Texas, this chapter will demonstrate how the valuation of land based on vulnerability to natural hazards was intertwined with policies of white supremacy. Flooding had a substantial role in shaping the changing historical geographies of race and class in the city during the era of Jim Crow, when state policy actively promoted anti-black racism to disenfranchise non-whites. One of the privileges that non-whites were denied was "staying dry." In Austin, however, it is important to note that Jim Crow also meant the subjugation, racialization, and devaluation of Mexican and Mexican-American residents, as well as African-Americans. In order to understand the ways in which non-white minorities were exposed to increased risk of flooding, it is crucial to understand the city's tri-racial form of segregation and valuation. In Texas, and throughout the Southwest, limiting our understanding of the powerful effects of being non-white to just African-Americans masks decades of injustices and ignores the crucial role that the "darkening" of Hispanics played in leaving them more vulnerable to environmental risk.

Introduction: Environmental Vulnerability and Racism

"Racism," Ruth Gilmore noted, "is the state-sanctioned and/or extralegal production and exploitation of group-differentiated vulnerability to premature death" (Gilmore, 2007, p. 247) . For Gilmore, difference is fatal. Somatic and morphological differences, in particular, mark people more or less susceptible to risks. Racism is ensured by violence and maintained because certain kinds of people, as a result of the bodies they have or allegedly have, are given a greater burden and excluded from sources and expressions of social power, that increase could relieve the costs associated with their vulnerability. State-sanctioned and extra-legal systems of white supremacy have ensured by tacit and explicit forms of repression that whites, regardless of class, were put in relatively less-vulnerable positions than African-Americans and other non-white minorities. Although true throughout American history it was particularly explicit in the American South during the period of de jure racial apartheid known as Jim Crow. By monopolizing political institutions through legal and extralegal means, whites were afforded better access to food, jobs, education, housing and, as a result, had better access to other forms of social and financial capital.

The historical geographies of city neighborhoods in the South, their patterns of settlement, immigration, and emigration are inseparable from the environmental injustices placed upon communities of color. From at least the early 1890s, non-whites in the South were increasingly excluded through law and custom from having access to infrastructural improvements such as sewage, drainage, and paved roads; they were often forced to settle in areas that had greater exposure to environmental hazards such as flooding and disease (Kellogg, 1977, p. 313; Haar and Fessler, 1986, pp. 1–2; Colten, 2005, p. 80). Additionally, it has repeatedly been shown that areas throughout the United States where non-white communities are or have been segregated have higher rates of exposure to air toxins, waste sites, and other artificial pollutants than white areas (Pulido, Sidawi et al., 1996; Bullard, 2000).

Scholars studying the social dimensions of the environmental hazards have pointed out how exposure to a risk has a social as well as an ecological component (Hewitt, 1983; Hewitt, 1997; Pelling, 2003). However, only recently, with the notable exception of Mike Davis' work, has the literature on environmental racism given attention to other natural hazards such as floods, tornadoes, fires, landslides, and earthquakes (Davis, 1998; Steinberg, 2000; Bullard, 2008). Attempts to address this oversight have been made most forcefully by Craig Colten (2005), but he understands racism to be a problem of attitudes that structure social relations and not systemic practices that generate group-differentiated exposure to vulnerability. Non-white minorities were made vulnerable because of two interrelated processes: (1) specific policy responses and systemic mechanisms of exclusion (such as local government policies of segregation, the financial costs of better homes, or private racial covenants) that restricted there access to better land; and (2) the actual material conditions among African-American and Mexican households (such as poverty, lack of education, poorer quality housing, etc.) that heightened their flood risk. This chapter addresses the shortcomings in the literature by more explicitly tying socially structured group-differentiated vulnerability to practices of racism, pre-existing natural hazard risks, and the historical evolution of neighborhood development.

Secondly, this chapter critiques David Delaney's work on what he has called the "spatiality of Jim Crow" or "the shifting spatiality of southern race relations, that is, an increasingly rigid form of segregation" for whites and non-whites in the American South from the latter portion of the 19th century to the middle portion of the 20th century (Delaney, 1998, p. 99). Although Delaney's discussion about changes in urban space do help to characterize the racial transformations in Austin's neighborhoods during this period, by focusing too narrowly on the biracial system, he does not adequately address the tri-racial system that existed throughout much of the Southwest, particularly in Texas. In states where legislation enforced a system of biracial apartheid and white supremacy over blacks, the position of Hispanics, Mexicans (particularly in Texas), within this hierarchy remained complicated. Legally, only African-Americans (keeping in mind the one-drop rule) were subjected to de jure forms of racial discrimination and disenfranchisement but most Hispanics, save those in the upper-class, suffered from *de facto* forms of exclusion and repression.[1] The system of racial exclusion in the Southwest, however, was not static and Hispanics, increasingly paying the price for being brown, were denied the privileges of being white granted to other immigrants (Sanchez, 1993; Rosales, 2000; Foley, 2004; Gomez, 2007).

We will argue that a pre-existing environmental hazard, namely periodic flooding, played a substantial role in shaping the racial geographies of Austin. Despite that Hispanics, African-Americans, and poor whites all settled in the most flood prone areas, Hispanics would have the greatest potential exposure to the flooding. As we will show, restricted from other areas of the city, when Hispanics migrated from their dominate enclave in the western portion of the downtown (the most vulnerable area for floods), they moved to the Eastside and into another very flood-prone area. This was not completely determined by the potential exposure to the natural hazard and show how the pattern of geographic migration was uneven and related to complex interactions between social and natural processes. For example, the most flood-prone area – the flatter lands along the north banks of

---

1   The one-drop rule is a standard for judging the black ancestry of a person that is unique only to the United States. Simply, the idea states that one drop of "African" ancestry makes a person black. The one-drop rule has a complex legal history that continues to be in use today, as courts upheld the validity of the one-drop rule when making designations for administrative purposes such as education or birth ancestry. While the one-drop might have existed in custom prior to the 1920s, it was during the Progressive Era that the one-drop rule became accepted throughout the United States as the perimeter for determining racial "purity" that separated blacks and whites Davis, F.J. (1991). Who is black?: one nation's definition. University Park, Pennsylvania State University Press.

the Colorado River – was the last place of Hispanic immigration on the Eastside because they were able to move into the area only after lower class whites finally left in the late 1940s and after much of the potential flood risk had been mitigated by the damming of the river (Austin).

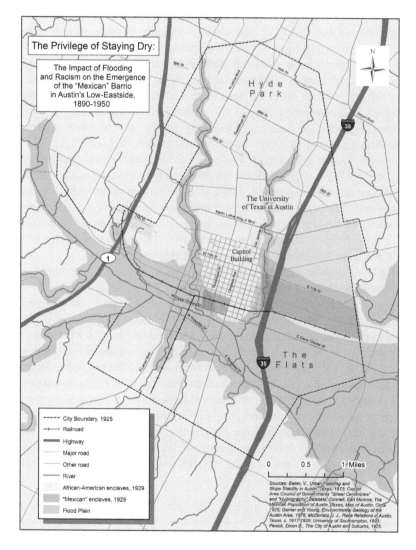

Figure 10.1   Austin, Texas. Source: Timothy Huynh, 2010

The History of Flooding in Austin

In Austin, periodic flooding is a persistent and significant environmental hazard. Like the Mississippi in New Orleans, the Colorado River, which flows through Austin, provides the city immeasurable resources such as fresh drinking water, irrigation for farms in surrounding communities, recreation, a line of transport, and a source of electrical power. With these enormous benefits come the significant drawback of recurring flooding along the river and its tributaries. Visible in the geological stratigraphy around the city and as documented since at least the early 1830s, Clay estimates that between 1900 and 1939 floods along the Colorado have cost over 85 million dollars and had killed at least 99 people of all races and classes (Brown 1820–36, pp. 43–44; Brown 1843, p. 3; Brown 1851–52, p. 34; Brown 1869, pp. 6–11; Brown 1870, pp. 29–33; Brown 1873, p. 36; Clay 1948, pp. 179–181).

In 1893, construction finished on the new "Austin Dam." Sixty feet tall (at the time it was the largest dam in the world spanning a major river), the dam was the first attempt to harness the river's power for electricity (Wilson, 1999). The Colorado River was known to flood – the flood of 1869 was so great it was thought to have changed the shape of the channel of the river; it inundated most of the city center and many communities downstream – but engineers thought the dam could withstand a major flood event (Brown 1869, p. 6–11; Taylor, 1910).

The flood on April 7, 1900, only seven years after the dam was built, proved them wrong. The flood was caused by a two-day storm in the High Plains halfway between Lubbock and Amarillo. The Colorado River, over-filled from the heavy storm, snapped the dam, sending "an immense volume of water" down the river, which was later estimated to be about 151,000 feet per second (Adams, 1990 p. 10). A wall of water at least 50 feet high flooded the city, causing in excess of nine million dollars in property damage, officially killing at least nine (eight were men working in the dam), and destroying the dam (Cline, 1900; Orum, 1987, p. 90; Wilson, 1999). There were reports that other white and black people drowned or disappeared but few were confirmed (*Dallas Morning News Sunday*, 1900; *Statesmen*, 1900). The damage to property along the Colorado's tributaries, Shoal and Waller Creeks, was especially severe. Along the banks of Waller Creek, the devastation went for several blocks and destroyed many houses mainly occupied by African-Americans. Rising more than 20 feet in less than two hours, Shoal Creek overflowed for at least a half mile and flooded "the houses situated along it banks for several blocks" (*Austin Daily Tribune*, 1900). At the time the largest Hispanic enclave was located along the banks of Shoal Creek in the city's downtown and it was reported that "a number of the Mexican population [had to move] to the highest point in the city" (*Statesmen*, 1900).

Another area that was extensively ruined was then known as "The Flats," which was the section of the city on the north bank of the river south of present-day E. Cesar Chavez Street. The area then was a predominantly white district of Germans and Swedes (Keahey 1976, p. 91). Mary Gordon, a young girl of German descent, described her predicament in a letter to her uncle written several days after the flood. Comparing the events to the Johnstown Flood of 1889, she describes the fear of being in the flood and the sense of loss over the people, and particularly the children, who were killed. When the flood began, Gordon recalled, everyone living in this area had to flee their homes for higher ground, and from her position there she watched three houses wash away. She moved, she wrote, to a different "portion of town" that was "level and white Sandy" because the Flats were "just awful in the rainy weather. The mud was so sticky and bad" (Gordon, 1900).

The dam was rebuilt in 1912 but was once again destroyed by a flood in April 1915 that reportedly caused over $650,000 in damages to buildings, roads, and bridges within the city (Orum, 1987, p. 91; Wilson, 1999). The majority of the flood damage and deaths were caused by flooding along two creeks, Shoal and Waller, which had turned into "raging torrents" (Bunnemeyer, 1915). Whole sections of the city were under water for hours, and houses built along the banks of Waller Creek were again destroyed. The flood was the deadliest in Austin's recorded history officially killing at least 35 people, who were mostly white people living near Waller Creek, but at one person also died along Shoal Creek (*American*, 1915; City of Austin, 1916).

Twenty years later, two significant floods occurred one year apart, in 1935 and 1936. The 1935 was the largest flood in recorded history. The peak discharge is estimated to be 481,000 cubic feet per second and the river reached the highest stage since the 1869 flood (City of Austin, 1976). The flood devastated many buildings and "some bridges [such as the Montopolis Bridge] that were ... considered to be above any possible high water" were also destroyed (Lowry, 1937 p. 13). The *Austin Statesman* reported, "Scores of persons living in the lowlands in East Austin were rendered homeless by the flood waters which drove them from their houses and which menaced property up to Third street on the east side of the city" (Wegg, 1935). It was estimated that the flood left up to 3,000 residents of East Austin homeless and most of their possessions destroyed by mud, especially those near the riverbank at East Avenue and Waller Creek (*Austin American*, 1935); (1800 Block of Holly St. 1935).

A little more than a year later, in September 1936, Austin was hit by yet another series of floods. A hurricane passing between Corpus Christi and Brownsville from September 17–23 inundated streams and was followed by another storm that flooded the lower section of the Colorado watershed. Before the water from the prior two storms receded, on September 28, a third and larger more storm came (Lowry, 1937, pp. 14–15). The river reached a height of 34 feet, and many of those living in low areas were evacuated as the Colorado flooded streets along the river's north bank (Long, 1936). With water advancing as much as a mile north of the riverbank, many streets and homes in East Austin, some were under as much as eight feet of standing water. Waller Creek too had backed up covering areas closer to the downtown, and substantial flooding also occurred in low-lying areas along the river's south bank.

**Figure 10.2   1800 block of Holly St., Austin, Texas, 1935. Courtesy of the
Austin History Center**

For the last 70 years six dams located upstream from downtown Austin have helped mitigate the risk of flooding by controlling the river's flow and while flooding continues to be a hazard along the Colorado River, the massive devastation of these floods and its impact on Austin has largely been forgotten. The dams all completed by 1951, most by early the 1940s, form a group of lakes, known as the Texas Highland Lakes and, in addition to drastically mitigating the flood risk, provide Austin with a stable source of water and electricity. A seventh and final dam, Longhorn Dam, was built downstream, just to the east of downtown Austin, and only was completed in 1960. Forming Lady Bird Lake, Longhorn Dam, unlike the other dams, does mitigate flood risk or provide fresh drinking water, but instead was built to create a reservoir that was necessary for a new power plant that was going to be constructed along the river, known as the Holly Street Power Plant. The damming of the Colorado upstream had reduced the ecological threat of flooding and made possible certain changes in urban environment but, other coterminous changes in the social environment, particularly the location of certain races and classes of people, also influenced how the urban environment in Austin's eastside could be transformed between 1900 and 1940. The following sections will explore these social changes and suggest that the interactions between race relations and flooding or the human alternation of the flood risk shaped how the urban environment developed.

The Racialization of Mexicans and Mexican-Americans in the American Southwest

In the United States, racial categories are products of history that have often served social purposes, dramatically influenced economic social relations, and developed very specific racial hierarchies (Goldfield, 1997). Ideas about race and racist practices had a significant influence on the creation of an economic underclass of non-white people that were an easily exploitable labor force. Throughout large sections of the country, African-Americans occupied this lower social position, but in the Southwest, the large number of Hispanic immigrants, particularly Mexicans, became a significant population of exploitable labor. Racial ideas in the United States infused all aspects of society and culture and strongly influenced a person's access to resources, ability to find work, right to join a union, social station, and, for foreigners, ability to become a citizen (Meeks, 2007, p. 17).

Despite the ease with which the economy accommodated Hispanic immigrants, socially they confounded the American bi-racial system. The presence of Mexican immigrants and citizens in the Southwest often created racial confusion among some white Americans, who, coming from other parts of the United States, had traditionally known a structured hierarchy of social relations built upon a racial binary. Mexicans were particularly difficult for white Americans to classify. In Mexico, the great amount of racial mixing was seen as an improvement of the population, but for many Americans, racial mixing was seen as a despicable act, tainting racial purity (Foley, 1997, pp. 57–58; Meeks, 2007, pp. 51–52). Mexicans were painted as inferior and less human due to both their history of racial "mixing" and to their having lived under "backward" Spanish and Mexican political systems. By the end of the 19th century, accelerating economic development facilitated the solidification of racial boundaries in the Southwest, with Hispanics more-or-less categorized as "non-white" persons with a distinct social, political, and economic status that denied them access to the privileges of whiteness (Rodríguez, 2005; Gomez, 2007).

In the prevailing racial social system during the turn of the 20th century, however, it would be more accurate to understand Mexicans in Southwest as occupying a "racial purgatory," subject to the stigmas and discriminatory practices faced by groups classified as non-white, without being legally defined as non-white. Unlike members of other nationalities, who, after one generation, were able to achieve status as Americans and sometimes even the benefits of being white, Mexicans were always labeled as unwanted "aliens," regardless of the time they had spent in the United States (Meeks, 2007, p. 115). Although "Mexican" refers to a nationality, rather than a separate race, many whites refused to accept this distinction, and anti-immigration activists (or nativists) often argued that nationality was not simply determined by birthplace but by race as well (Foley, 1997, p. 57; Meeks, 2007, p. 114). While Hispanics were denied the privileges of being white, legally they were not black. In Texas, for instance, Hispanics were considered white because blacks were defined as only those people with

African ancestry and only those people were subject to the legal force of apartheid regulations.

In practice, however, Mexican and Mexican-Americans were not considered a part of this non-black group and were often subjected throughout the Southwest to intense forms of *de facto* segregation and discrimination (Foley, 1997, pp. 40–63). Hispanics had to contend with widespread residential segregation, segregated schools and public facilities, and voting rights obstacles (Montejano, 1987, pp. 167–169; Meeks, 2007, p. 160). Like other non-whites, Hispanics also were subject to other forms of extra-legal violence. In 1922, at the height of nativist campaigns, 22 Mexicans were lynched in the United States, and at least half were lynched in the state of Texas (McDonald, 1993, p. 33). Furthermore, throughout the region, Hispanics were relegated to the lowest economic station and their resulting poverty influenced their social status and the neighborhoods in which they could live (García, 1981).

The Emergence of a Tri-racial Community in Austin

Any natural processes interact in complex ways with social structures, reshaping those social structures and transforming how natural processes are understood (Harvey, 1996, Chapter 6). The conjunction of environmental and racial factors can been seen in an advertisement for one of Austin's then-suburban neighborhoods, Hyde Park, which developed between 1890 and 1930. Initially designed for wealthier people by the developer, the neighborhood ultimately became a working-class area consisting of small bungalows that was, significantly, advertised as "exclusively for white people" (Advertisement for Hyde Park Circa 1900). Moreover, advertisements after 1900 noted that the neighborhood was "free from mud and dirt" and "185 feet above the river," which was far above the floodplain. Although some blacks lived in the neighborhood as servants, with different sorts of housing arrangements, they were prohibited by restrictive covenant from owning any property (Sitton, 1991, p. 107). Hispanics, while not legally barred from owning property in the area, appear to be mostly absent. For instance, Sitton (1991) documented how many residents in Hyde Park were non-Texans, and a large number were foreign born northern Europeans, but she does not mention a single Hispanic household.

By 1930, about the time that the construction of Hyde Park was completed, Hispanics had a considerably sized community, probably numbering around 5,000, or about 9 percent of Austin's population (McDonald, 2006, p. 132).[2] Like most Southern cities, before 1917 Austin was primarily a biracial city, but by the 1930s, a tri-racial urban space of white, black, and brown peoples emerged. A

---

2   In the early 20th century, Austin's Hispanic population was composed of Tejanos and newer immigrants from Mexico, which we will, following the usage at the time, refer to them only as Mexicans.

HYDE PARK

The most beautiful, healthful, and practical place for homes in the city of Austin. It's the safest place for investment. The terms offered are remarkably easy. The prices are very reasonable. Any person buying two lots WILL BE GIVEN ONE LOT FREE OF COST. There are six miles of beautiful graded streets in HYDE PARK, and a magnificent

SPEEDWAY FROM THE PARK TO THE CITY.
THE FINEST DRIVE IN TEXAS.

HYDE PARK IS EXCLUSIVELY FOR WHITE PEOPLE.

The main line of Electric Street Cars run into and around a belt in the Park. Free Mail Delivery twice a day. There is no limestone dust. The soil is the best for Fruits, Flowers and Lawns. No one thinks of taking a carriage drive without going to Hyde Park. The drives are free from mud and dust. The scenery is interesting. The altitude of Hyde Park is 185 feet above the river. Hyde Park is Cool, Clean and Restful. Invest while YOU CAN SELECT, and SECURE ONE LOT FREE. If you wish to buy on the installment plan the terms are $3.00 per month on each lot. If you pay all cash a discount of 8 per cent will be allowed. If you wish to invest and do not live in Austin, we will pay your fare both ways, if the distance is not over 300 miles. Strangers who wish to see the city can have a Free Carriage by calling at our office.

Extraordinary Inducements Are Offered

To persons who will agree to erect good houses. If parties wish to build in Hyde Park we will trade lots for other Austin property on a fair basis, and DONATE ONE LOT as a Premium. Beautiful Views of Hyde Park, and of THE SPEEDWAY sent free upon application. Write to us, or call at 721 CONGRESS AVENUE, AUSTIN, TEXAS.

M. K. & T. LAND AND TOWN CO.
M. M. SHIPE, General Manager

Figure 10.3   Advertisement for Hyde Park. Courtesy Austin History Center

sizable population of African-Americans, mostly ex-slaves, and a much smaller Mexican community lived in Austin at the turn of the 20th century, but by the 1930s, the situation had changed drastically, as the Hispanic population had grown significantly, primarily from immigration occurring between 1917 and 1929 (Martin, 1933). Between 1900 and 1930 over 10 percent of Mexico's total population emigrated, fueled, in part, by internal unrest, economic dislocation, and decades of instability resulting from the Mexican Revolution (McDonald, 1993, p. 15). Moreover, during World War I, increases in production in Northern industrial areas led to increased demand for labor, and because white males were mobilized for war and legislation restricted immigration from Europe and Asia, employers drew upon the South's large population of blacks, creating a labor shortage in many Southern states. In order to solve the problems of labor scarcity, especially for agriculture, Texans turned to Mexican immigrants (Foley, 2004, pp. 40–63). Therefore, the migration of African-Americans and Hispanics was interrelated and occurred in two phases: (1) between 1916–1920, large numbers of blacks emigrated north, while Mexican immigration was not as noticeable by comparison; and (2) between 1921–1929, black emigration slowed, while

Mexicans immigration increased rapidly (McDonald, 1993, p. 66) McDonald estimated that the Black population between 1910 and 1920 fell from about 7,500 to about 7000, while the numbers of Mexican immigrants increased from around 500 to 5000 over the same period (McDonald, 1993, pp. 462–463).

Prior to 1920, Mexican immigration was overshadowed by the Great Migration and not considered serious, as the common belief was that Mexicans would eventually return to Mexico. While some did return during World War I, mostly out of the fear of conscription, the numbers leaving were outpaced by new arrivals (McDonald, 1993, pp. 68, 70–71). As the volume of Mexican immigration increased, so did opposition to their presence. In the 1920, people more and more advocated new laws to restrict Mexican immigration such as imposing quotas (McDonald, 1993, pp. 78–79; Foley, 1997, pp. 51–53). Support for these proposals often came from unions or organized labor groups worried about the impacts of immigration on wages and from the Ku Klux Klan. In response to political pressure, in 1929 Hispanics began being deported more frequently, often despite proof of legal residence (McDonald, 1993, pp. 80–83).

Notwithstanding efforts to remove Hispanics, particularly poorer ones, they remained in Texas. Working many of the same jobs that had been previously reserved for blacks, particularly in agriculture, Hispanics were confined to certain sectors and low-wage jobs, ensuring their increased subordination and undesirable status as non-whites. While the racial category "Mexican" was already used colloquially, it became codified in 1930, when "Mexican" appeared in that year's US census (McDonald, 1993, p. 99). The category, removed from later censuses, seems to mark a turning point, as it reflects recognition by the government that the Southwest was officially a tri-racial system.

The Tri-racial Segregation of Austin

Between 1900 and 1940 racial demographics and housing patterns in Austin changed significantly, seemingly the result of the implementation of public and private means of racial segregation. Southern whites tried to enforce the subordination and segregation of blacks after World War I through both legal and extralegal means, but there were no constitutionally sanctioned separate-but-equal standards for Hispanics. Despite this, racist practices would be extended to include Hispanics, as ultimately, black and brown peoples were seen as "different aspects of the 'same' race problem" (Montejano, 1987, p. 262). The pattern of racial discrimination was also spatial: while the two non-white groups did not occupy the same blocks, by the mid-20th century their two most dominant enclaves were located in the area known as East Austin, a section of the city east of East Avenue (the present location of IH-35) between 19th Street (now Martin Luther King Blvd.) and the Colorado River (Jackson, 1979, pp. 70–75).

African-Americans had been in East Austin as early as the late 1800s, and by the 1880s African-Americans, as in other Southern cities, were scattered across

Austin, with small communities in virtually every neighborhood, including several suburban areas (Kellogg, 1977, pp. 310–311). By 1910, however, African Americans had been driven by a number of push-and-pull factors to congregate in larger numbers on Austin's Eastside, and by 1930, the largest concentration of Austin's African-Americans were living within this single area (Human Relations Commission, 1979, p. 11). Containing its own businesses, churches, and social communities, East Austin became the city's African-American enclave (Mears, 2009, pp. 148–150). Certainly, as other research has suggested, African-Americans would have found their own neighborhoods more appealing, in part because of their relative distance from the discrimination and social violence of whites (Smith and Woodward, 2002, pp. 20–21). However, as in other municipalities in the South, by the 1920s many of Austin's newer neighborhoods had instituted deed restrictions, preserving their communities "for whites only," and prohibiting "people of African ancestry" from buying or renting homes almost anywhere in the city other than East Austin (Humphrey, 1997, p. 36; Klarman, 2004, pp. 142–146).

Municipal planning also had an important hand in shaping the racial patterns of settlement. In 1928, the City of Austin hired Koch and Fowler, two engineers, to develop an urban plan. While most of the report concerned paving roads and park construction, a section explicitly addressed the "race segregation problem" (Koch and Fowler, 1928, p. 57). The problem, as they understood it, was how to constitutionally segregate minority communities. In 1917, the Supreme Court held in Buchanan v. Warley that using residential zoning codes to enforce the racial segregation of a city placed an unconstitutional infringement on private land owners to dispose their property, and as a result municipalities had to search for alternative methods to impose the racial segregation of neighborhoods (Klarman, 2004, pp. 79–85, 90–93). Koch and Fowler offered a method for segregating races that was legal: they would create "a Negro district; and that all facilities and conveniences [would] be provided the Negro in this district as an incentive to draw the Negro population to this area" (Koch and Fowler, 1928, p. 57). While the plan mentioned that African-Americans resided throughout Austin, the authors emphasized that there was already a high concentration of blacks in East Austin and suggested this should be a site for a "Negro district." The plan would save the municipality money by avoiding duplication of other segregated facilities such as schools, parks, etc. and would also encourage the spatial separation of the black and white people. The plan was adopted by the City Council, and soon after, Rosewood Park and Anderson High School were built in the district as the only park and high school "for Coloreds" in the city (Orum, 1987, p. 176). Of equal significance, was the extension of sewer service to African-American homes in the East Austin area by 1930, but its denial to other black enclaves in the city (Kraus, 1973, pp. 150–152).

While the Koch and Fowler report did not mention Hispanics as a part of the race segregation problem, by the 1930s, Hispanics were also increasingly concentrated in East Austin. Before the 1930s, Hispanics had been largely confined to a neighborhood located west of downtown's main retail businesses, close to the confluence of the Colorado River and Shoal Creek. There were probably a number of reasons for the movement out of the downtown and into east Austin. Connell, at the time, suggested, "Mexicans" found land was cheaper, no railroads passed directly through the neighborhood, no businesses or factories were located within the neighborhood, and residents had a greater sense of security, particularly "in the wake of business expansion" in the central business district (Connell, 1925, p. 4). Additionally, as in other Texas cities, it is important to emphasize that Hispanics in Austin had a limited housing choices due to the financial costs and racial covenants that barred sales to "non-Caucasians" and their access to public services was restricted to their own separate schools, parks, playgrounds, public housing projects, et cetera (Montejano, 1987, p. 265; Humphrey, 1997, p. 42; McDonald, 2005, p. 4).

Nevertheless, Hispanics appear to not have been as segregated from whites as blacks had been, at least in the earlier portion of the 20th century. The neighborhood of Hispanics immigration in East Austin had, for some time, been an area with a high concentration of poorer white communities, mainly Swedes and Germans. Between 1915 and 1950, the racial relationship between whites and Hispanics would change dramatically and is demonstrated in the history of two elementary schools, Metz and Zavala. Metz Elementary, from the time of its founding in 1916, was composed primarily of white students, but by 1926, 20 percent of the student body was Hispanic Montejano, 1987, p. 262). Following an the increase of Hispanics into the area, Zavala Elementary, a separate school for the Hispanic children that had attended Metz, was built in 1936, just four blocks north of Metz. While some Hispanic students stayed at Metz, most went to Zavala, and by 1940 Metz was 90 percent white Montejano, 1987, pp. 265–267). The increasing white exclusivity of Metz was short lived, however, only a decade later, the number of Mexican students enrolled at Metz began to rise dramatically, reflecting both the population growth of Hispanics in the neighborhood and the rapid exit of whites.

Housing Conditions and the Impact of Flooding on the Settlement Patterns of African-Americans and Hispanics

The development of East Austin was greatly influenced by the recurring flooding of Colorado River and Waller Creek. A hill between 7th and 11th streets is the highest ground in East Austin, atop of which stands the historic French Legation building. When the area was first being developed, the land around the hill "away from the Colorado's frequent floods sold first" (Keahey, 1976, p. 91). The area below the hill was slower to develop, but it gradually by 1891 became one of Austin's first suburbs, mostly occupied by German and Swedish immigrants. "Despite periodic flooding all the way to 6th street," Keahey notes, "parts became quite prestigious places to live," but the 1912 plans to beautify the area with parks, promenades, mansion rows with fountains, and terraces never materialized because of "disastrous floods in … 1915" (Keahey, 1976, pp. 91–92). Nevertheless, development in the area continued throughout the 20th century and it underwent substantial demographic changes. Increasingly Hispanics would come live on the south side of the hill, while African-Americans would congregate on higher ground to the north.

In addition to private racially restrictive covenants and the city's policies of segregating social services, monetary price would have significantly influence on the mobility of African-Americans because irrespective of the neighborhood, during the early portion of the 20th century, in Austin they lived in very poor, overcrowded, and dilapidated houses or shacks. However, when considering the other options, East Austin had the best land at the lowest financial cost because at least it was outside the floodplain. In the few areas of the city where African-Americans lived near whites, they lived in back houses or in the footprint of several former "freemen" (former slaves) colonies that had formed throughout Texas after the Civil War. However, these settlements were either far from the city or were located along the flood-prone creek banks, such as Waller and Shoal, which were also open sewers (Hamilton, 1915, p. 7 and 48; Manaster, 1986, p. 55; Mears, 2009, p. 155).

While Hispanics were not as restricted by private racially covenants as African-Americans, like blacks, they were severely limited in their geographical mobility by the price land and social prejudice, and throughout Austin their housing conditions were poor and overcrowding was the norm. "No matter where we find Mexicans they occupy about the same type of physical surroundings," Connell noted. "They live in the poorest houses, near the railroads, with no paved streets or sidewalks, in the business slum districts or on the creeks" (Connell, 1925, p. 6). From at least the 1850s, a Hispanic enclave was located at the mouth of Shoal Creek. Known for being lined with trash, manure, and raw sewage because the city dump was located in a ravine that emptied directly into the creek, and the area around the creek's mouth also bordered the "tenderloin" district, so that many Hispanic family homes were located adjacent to "houses of prostitution" (Hamilton, 1915 pp. 6 and 57–60). Associated with filth, crime, and a general immorality, "Mexico," as neighborhood was called at the time, was also one the most dangerous flood-prone areas, and rain would have created very salient fears among the general population because property damage and loss of life could often be very significant.

The movement of Hispanics around Austin cannot be separated from the proliferation of private deed restrictions and the city's plans to create a "Negro District," which, it appears, Hispanics were largely thought to belong in or near. By the late 1920s, in part because of the deplorable conditions downtown, Hispanics began migrating to East Austin in greater numbers. Despite being almost two miles from the city center and not having paved roads or sidewalks, the neighborhood was a significant improvement over the Shoal Creek area because it was laid out in blocks with named streets and was not in the floodplain (Connell, 1925, p. 4). However, the 1928 master plan, which had outlined a blueprint for racial segregation, also had zoned much of East Austin, particularly the main thoroughfare in the newly emerging Hispanic enclave, for industrial and heavy commercial uses. While Hispanics moving into the area in the 1930s were probably largely unaware of these zoning rules and might have been eager to escape the deleterious effects of pollution from factories located downtown, by the mid-1940s most of Austin's light industrial firms would be located their neighborhood.

Flood patterns also continued to influence where Hispanics settled. The area where the newly forming Hispanic enclave in East Austin was once described as "most heterogeneous area of the city" because the largest African-Americans enclave bordered to north, and a poorer community of "unskilled and semi-skilled whites" was located to the south (University of Texas. Bureau of Research in the Social Sciences, 1941, pp. 21–22). However, by 1950, nearly all whites had exited the area, perhaps wishing to escape the new industrialization, older and declining housing stock, or the presence of minorities in East Austin. Hispanics, restricted from other areas, rapidly moved into the area of white flight, which was the area that most susceptible to the flooding of the Colorado River, known as The Flats, and already on its way to becoming another slum. Although by 1945, the construction of large dams upstream had mitigated the risk from floods in this area, and perhaps this led some white families in the area to hold on to their homes longer and even fight to preserve the area as an all-white neighborhood (as the history of the Metz School showed), the neighborhood had been in a state of decline since the 1915 flood, which the floods in the 1930s only hastened.

Conclusion

East Austin developed as a lower-class ghetto, in part, because of the ways that Austin's African-Americans, Hispanics, and poor whites were segregated and subjected to persistent flood hazards. Prior to the solidification of East Austin as the main zone for people of color, the only neighborhoods available to African-Americans and Hispanics close to downtown were flood-prone areas where deaths and damage were frequent. These areas had the lowest land values, and those who had the financial means to live elsewhere did, but prejudice also had a significant hand in affecting how risk was distributed. The privilege of living in safer areas was largely afforded only to whites, who could pay the financial price to live on safer land and did not have to pay the social price of being brown or black. It was not only class but racism that drove the relationship between flood risks and neighborhood demographics. Poor whites had occupied the floodplain but as the spatialities of Jim Crow became more pronounced, so too did the benefits of being white, which allow certain kinds of people the opportunity to enter into select spaces of the city that would be less vulnerable to flooding and the long term negative consequences on the quality of houses from being located in the floodplain.

Patterns of racial hierarchy and discrimination in Austin were similar to those throughout the Southwest, where non-whites were disenfranchised by a system of racial apartheid that help transfer costs of being vulnerable to hazards away from whites. Placing a greater burden upon other social groups, institutions of white supremacy were designed to ensure whites were always dominating non-whites. Excluded from local government, education, and job opportunities that help to create sources of social power, non-whites were always at a disadvantage and lacked strong means to control their collective futures. Whether it was zoning regulations, limitations on public services, or how changes in the flood risk would affect their neighborhoods, decisions about changes to East Austin always came from outside. While African-Americans were certainly the victims of white domination, in Austin, Hispanics paid the highest price from flooding. Perhaps this was due to their recent arrival and lack of historical memory, linguistic barriers, or the lack of political organization. Whatever the reason, Hispanics remained the most vulnerable to the environmental threat of flooding and its consequences on the quality of life in their neighborhoods.

References

1800 Block of Holly St. 1935. PICA 21147, Austin History Center, Austin Public Library.

Adams, J.A. (1990). *Damming the Colorado: The Rise of the Lower Colorado River Authority, 1933–1939*. College Station, Texas A&M University Press.

Advertisement for Hyde Park Circa 1900. PICA 25419c, Austin History Center, Austin Public Library.

Austin. map. Made by Timothy Huynh, 2010.

Austin American (1915). 35 persons are believed drowned. *Austin American*.

Austin American (1935). Flood crest passes Bastrop and Smithville. *The Austin American-Statesman*.

Austin Daily Tribune (1900). Repetition of the Johnstown flood. *Austin Daily Tribune*.

Brown, F. (1820–36). *Annals of Travis County and of the City of Austin: From the Earliest Times to the Close of 1875*.

Brown, F. (1843). *Annals of Travis County and of the City of Austin: From the Earliest Times to the Close of 1875*.

Brown, F. (1851–52). *Annals of Travis County and of the City of Austin: From the Earliest Times to the Close of 1875*.

Brown, F. (1869). *Annals of Travis County and of the City of Austin: From the Earliest Times to the Close of 1875*.

Brown, F. (1870). *Annals of Travis County and of the City of Austin: From the Earliest Times to the Close of 1875*.

Brown, F. (1873). *Annals of Travis County and of the City of Austin: From the Earliest Times to the Close of 1875*.

Bullard, R.D. (2000). *Dumping in Dixie: Race, Class, and Environmental Quality*. Boulder, Westview Press.

Bullard, R.D. (2008). "Differential vulnerabilities: environmental and economic inequality and government response to unnatural disasters." *Social Research* 75(3) 753–784.

Bunnemeyer, B. (1915). "Floods in Texas during April and May, 1915." *Monthly Weather Review*, 186–189.

City of Austin (1916). City Council Ordinances, April 13.

City of Austin (1976). Flood plain information Colorado River-Onion Creek to Montopolis Bridge. Austin, Army Corps of Engineers.

Clay, C. (1948). *The Lower Colorado River Authority: A Study in Politics and Public Administration*. Austin, University of Texas.

Cline, I.M. (1900). "Special report on the flood in the Colorado Valley, Texas, April, 7–17 1900 and other floods during the same period." *Monthly Weather Review*, 146–150.

Colten, C.E. (2005). *An Unnatural Metropolis: Wresting New Orleans from Nature*. Baton Rouge, Louisiana State University Press.

Connell, E.M. (1925). *The Mexican Population of Austin*. Texas, University of Texas.

Dallas Morning News Sunday (1900). Death rides the torrent: eighteen lives lost by a mighty rush of water which sweeps away Austin dam.

Davis, F.J. (1991). *Who is Black?: One Nation's Definition*. University Park, Pennsylvania State University Press.

Davis, M. (1998). *Ecology of Fear: Los Angeles and the Imagination of Disaster*. New York, Metropolitan Books.

Delaney, D. (1998). *Race, Place, and the Law, 1836–1948*. Austin, University of Texas Press.

Foley, N. (1997). *The White Scourge: Mexicans, Blacks, and Poor Whites in Texas Cotton Culture*. Berkeley, University of California Press.

Foley, N. (2004). *Party Colored or Other White. Beyond Black & White: Race, Ethnicity, and Gender in the U.S. South and Southwest*. S. Cole, A.M. Parker, and L.F. Edwards. College Station, Texas A & M University Press: 123–144.

Franklin, R.W. (1935). From a great flood (negro spiritual). Austin History Center, AF W1700.

García, M.T. (1981). *Desert Immigrants: The Mexicans of El Paso, 1880–1920*. New Haven, Yale University Press.

Gilmore, R.W. (2007). *Golden Gulag: Prisons, Surplus, Crisis, and Opposition in Globalizing California*. Berkeley, University of California Press.

Goldfield, M. (1997). *The Color of Politics: Race and the Mainsprings of American Politics*. New York, New Press.

Gomez, L.E. (2007). *Manifest Destinies: The Making of the Mexican American Race*. New York, New York University.

Gordon, M. (1900). Mary Gordon papers. Austin History Center.

Haar, C.M. and D.W. Fessler (1986). *The Wrong Side of the Tracks: A Revolutionary Discovery of the Common Law Tradition of Fairness in the Struggle Against Inequality*. New York, Simon and Schuster.

Hamilton, W.B. (1915). A student survey of Austin, Texas: a digest of University of Texas bulletin no. 273. Austin, University of Texas.

Harvey, D. (1996). *Justice, Nature, and the Geography of Difference*. Cambridge; Oxford, Blackwell Publishers.

Hewitt, K. (1983). *The Idea of Calamity in a Technocratic Age. Interpretations of Calamity from the Viewpoint of Human Ecology*. Boston, Allen & Unwin: 1–32.

Hewitt, K. (1997). *Regions of Risk: A Geographical Introduction to Disasters*. Harlow, Longman.

Human Relations Commission (1979). Housing patterns study of Austin, Texas: a report. Austin, The Commission.

Humphrey, D.C. (1997). *Austin: A History of the Capital City*. Austin, Texas State Historical Association, in cooperation with the Center for Studies in Texas History at the University of Texas.

Jackson, R.E. (1979). *East Austin: A Socio-historical View of a Segregated Community*. Austin, University of Texas at Austin.

Keahey, K. (1976). *Austin and its Architecture*. Austin, American Institute of Architects, Austin Chapter and Women's Architectural League.

Kellogg, J. (1977). "Negro urban clusters in the Postbellum South." *American Geographical Society* 67(3) 310–321.

Klarman, M.J. (2004). *From Jim Crow to Civil Rights: The Supreme Court and the Struggle for Racial Equality*. New York, Oxford University Press.

Koch and Fowler (1928). *A City Plan for Austin, Texas*. Department of Planning.

Kraus, S.J. (1973). *Water, Sewers and Streets: The Acquisition of Public Utilities in Austin, Texas, 1875–1930*. Austin, University of Texas at Austin.

Long, S. (1936). River expected to go to height of 34 feet here – amply warned, many moved from low areas to safety. *Austin American*.

Lowry, R.L. (1937). *Flood Control* by Marshall Ford, U.S. Department of the Interior, Bureau of Reclamation.

Manaster, J. (1986). *The Ethnic Geography of Austin, Texas, 1875–1910*. University of Texas at Austin, 1986.: iii, 169 leaves.

Martin, R.C. (1933). "The municipal electorate." *Southwestern Social Science Quarterly* 14(3) 193–237.

McDonald, J.J. (1993). *Race Relations in Austin, Texas, c. 1917–1929*. University of Southampton.

McDonald, J.J. (2005). "From biparite to triparite society: demographic change and realignments in ethnic stratification in Austin, Texas, 1910–1930." *Patterns of Prejudice* 39(1) 1–24.

McDonald, J.J. (2006). "Marginalising the marginlised in wartime." *Journal of Ethnic and Migration Studies* 32(1) 129–144.

Mears, M.M. (2009). *And Grace Will Lead me Home: African American Freedmen Communities of Austin, Texas, 1865–1928*. Lubbock, Texas Tech University Press.

Meeks, E.V. (2007). *Border Citizens: The Making of Indians, Mexicans, and Anglos in Arizona*. Austin, University of Texas Press.

Montejano, D. (1987). *Anglos and Mexicans in the Making of Texas, 1836–1986*. Austin, University of Texas Press.

Orum, A.M. (1987). Power, money and the people: the making of modern Austin. Austin, *Texas Monthly Press*.

Pelling, M. (2003). *The Vulnerability of Cities: Natural Disasters and Social Resilience*. London, Earthscan Publications.

Pulido, L. and S. Sidawi, and Vos, R. (1996). "An archaeology of environmental racism in Los Angeles." *Urban Geography* 17(5) 419–439.

Rodríguez, V.M. (2005). "The racialization of Mexican Americans and Puerto Ricans: 1890s–1930s." *CENTRO Journal* 17(1) 70–105.

Rosales, R. (2000). *The Illusion of Inclusion: The Untold Political Story of San Antonio*. Austin, University of Texas Press.

Sanchez, G.J. (1993). *Becoming Mexican American: Ethnicity, Culture, and Identity in Chicano Los Angeles, 1900–1945*. New York, Oxford University Press.

Sitton, S. (1991). *Austin's Hyde Park: The First Fifty Years, 1891–1941*. Lubbock, Pecan Press Publications.

Smith, J.D. and C.V. Woodward (2002). *When Did Southern Segregation Begin: Readings*. New York, Palgrave.

Statesmen, A.S.-W. (1900). Austin's great loss of life and property. *Austin Semi-weekly Statesman*.

Steinberg, T. (2000). *Acts of God: The Unnatural History of Natural Disaster in America*. New York, Oxford University Press.

Taylor, T.U. (1910). *The Austin Dam*. Austin, University of Texas.

University of Texas. Bureau of Research in the Social Sciences. (1941). Population mobility in Austin, Texas, 1929–1931. Austin, The University of Texas.

Wegg, W. (1935). Persons caught in flood take to trees roofs. *Sunday American-Statesman*.

Wilson, J. (1999). The day the dam broke. *Austin-American Statesman*.

Epilogue

Sarah Dooling and Gregory Simon

This volume has traversed several themes related to urban vulnerabilities, their productions, and the material, ecological, economic, political, and cultural consequences of being made vulnerable. Each chapter has, in its own way, expanded conceptualizations of vulnerability from a point-in-time assessment and a static condition to a dynamic, multi-scalar historical process driven by context-specific interactions. Urban vulnerabilities are produced by geographies of wealth and risk accumulation governed by neoliberal policies and resource management practices in relation to urban fires and city parks; through the displacement of risk and the influence of disaster capitalism enabled by global north-south institutional arrangements; through sustainability planning efforts and environmental movements which concentrate benefits for the least vulnerable populations while exacerbating and creating new risks for already vulnerable populations; through activism that calls attention to future vulnerabilities associated with climate change, yet places activists at risk to psychological and physical harm; and through the confluence of segregationist politics and urban land economics that maintain risk to natural hazards for minority populations. Responses to the production of urban vulnerabilities – whether in the form of urban policy strategies intended to re-frame vulnerable conditions and re-purpose vulnerable urban spaces or in the form of political mobilization among historically marginalized and disempowered groups of women – point to the opportunities and potentials for resisting and even reversing vulnerable conditions and their associated processes of production. Capacity for resistance and reversing the production processes can be framed as a form of resilience, another theme explored by two contributors.

Contributors have drawn from a wide range of theories through which to conduct their analyses – including environmental justice, theories of feminist politics and identities, urban political ecology, social production of space, and environmental racism. The application of these theoretical lenses yields nuanced and more complex understandings of what it means for people and cities to be vulnerable; how conditions and perceptions of vulnerability can be manipulated and exploited to serve powerful urban elites; how processes of production can accumulate over time while also fluctuating in terms of levels and duration of risk exposure; how multiple processes and conditions of vulnerabilities can interact and influence each other across space and time; how efforts intended to enhance urban sustainability may actually deepen inequities and conditions of being vulnerable for already vulnerable groups of people, and in some cases facilitate co-occurring ecological vulnerabilities; and how communities can mobilize to transform vulnerabilities. The diversity of these theoretical frames demonstrates how developing more robust conceptualizations of vulnerability, as both condition and process, can inform empirical studies that can provide insight into the complexities of cities, nature and development.

Emerging from the theoretical and empirical contributions of this volume are three important trajectories for future research:

1. *Urban sustainability as a producer of vulnerabilities.* Many authors revealed the contradictions – and resulting social and ecological vulnerabilities – associated with efforts and movements intended to improve urban ecological function. These findings point to the need for future analyses of how economic and political conditions that undergird urban sustainability efforts may create new vulnerabilities, deepen existing vulnerabilities and perhaps conceal potential for future vulnerabilities. The dominant conceptualization of urban sustainability is one of maximum integration of ecological, economic and equity-related components of cities – the axiomatic triple bottom line. Yet, contributors to this volume have challenged this idea by demonstrating how urban sustainability efforts and plans result in uneven access to benefits with concomitant uneven experiences of being at risk. As discourses and practices of urban sustainability continue to be incorporated into urban planning and policies, it is critical to understand the unanticipated consequences of these policies and plans (e.g., unforeseen vulnerabilities that are produced), and how these plans and policies perhaps contribute to productions of vulnerabilities.

2. *Vulnerability and resilience.* A handful of authors in this volume demonstrate how conditions of vulnerability can be resisted and occasionally reversed. Resistance to and reversal of vulnerabilities can be understood as a form of resilience. Contributors in this volume operationalize resilience as engaging in political mobilization and community efforts to reduce environmental degradation. Resilience has gained traction as a concept related to building capacities to respond to an external perturbation without disruption in system functioning. Increasingly, resilience is associated with concepts of sustainability, whereby sustainability related plans and policies that integrate social, ecological and economic issues will lead to the development of resilient cities capable of responding to phenomena, for example climate change, without experiencing catastrophe. It therefore becomes important to understand the relationship between the production of urban vulnerabilities – as condition and process – and strategies for resistance and reversal in the context of neoliberal regimes, entrenched urban politics, asymmetrical power, and deepening inequities across lines of race, gender and income. Identifying possibilities for contesting and overturning vulnerabilities will, we believe, lead to more nuanced and robust conceptualizations of resilience, and its position within the discourse of sustainability.

3. *Urban vulnerabilities and climate change.* Climate change is understood by many scholars to be the most significant threat to the future of cities. With the failure of international governments to develop overarching climate action policies and commitments to reducing greenhouse gas emissions, cities are quickly becoming vanguards for the development of effective local plans and strategies to adapt to and mitigate future climatic conditions. Many of the early climate change strategies addressed biophysical and ecological impacts. More recently, increasing attention is being paid to the cultural and ethical implications of displacement and the proliferation of climate refugees; economic and cultural losses associated with climatic impacts on, and reduced access to, livelihood resources; and the burden of fiscal costs associated with adaptation and mitigation strategies for developing countries that are not significant sources of anthropogenic climate change. Many contributors in this volume detail how, in attempts to address ecological concerns, current conditions of vulnerability can deepen and persist over time when analyzed in the context of neoliberal policy. We anticipate that applying core urban vulnerability themes in this volume to the issue of climate change will elucidate linkages and feedbacks between the biophysical, social and ethical dimensions of global climatic change.

We hope that the work in this volume has provided conceptual and empirical contributions to urban vulnerability studies. We believe that this collection of research reveals important aspects of urban vulnerability – related to scalar interactions, regressive momentum, politicization and governance, contradictions and unanticipated consequences, and capacity for resistance and reversal – that call for further investigation. In many ways, the future of cities and the possibility for economically vibrant, culturally diverse, ecologically healthy, and just urban places rests upon critical engagement with the concept of urban vulnerability and its complexities.

Index